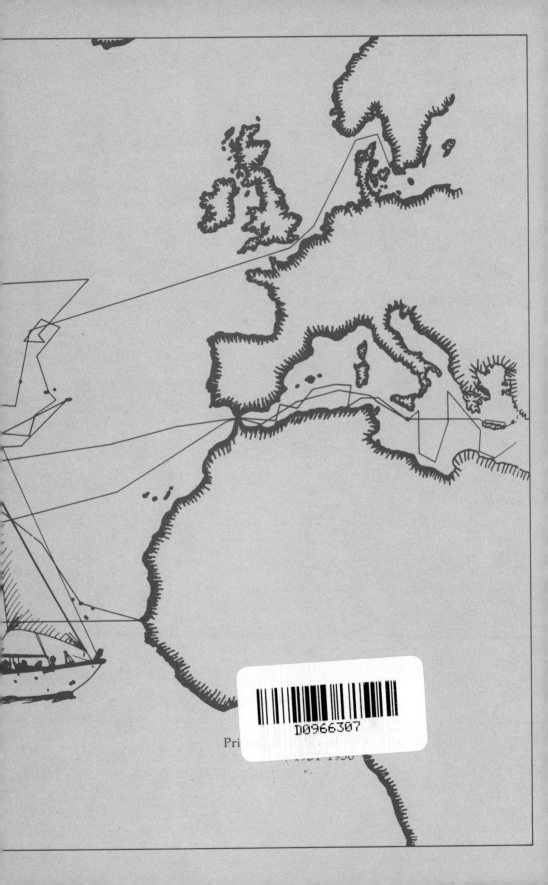

ON ALMOST ANY WIND

The research vessel *Atlantis*. "Catches [the original word for 'ketch'], being short and round built, be very apt to turn up and down, and useful to go to and fro, and to carry messages between ship and shore almost with any wind" (Glanville, 1625). *(Courtesy Woods Hole Oceanographic Institution.)*

ON ALMOST ANY WIND

The Saga of the Oceanographic
Research Vessel *Atlantis*

SUSAN SCHLEE

Cornell University Press / ITHACA AND LONDON

GC
57
534

For Tom Kelley and Mary Sears

> We old men are old chronicles, and when our tongues go they are not clocks to tell only the time present, but large books unclasped; and our speeches, like leaves turned over and over, discover wonders that are long since past.
>
> — Elizabeth Reynard, *The Narrow Land*

Contents

Acknowledgments

I am more surprised by receiving a gift from a stranger than by getting hours of work from a friend and more grateful for last month's offer of old photographs than for the stories told to me three years ago. Therefore, I am afraid that my efforts to thank all who have contributed to this account of *Atlantis* will be somewhat distorted. Still, I remember with pleasure that because Sheila Paine enjoyed my first *Atlantis* stories so much and because Ad Hause on the West Coast and Bill Doherty at my daughter's nursery school encouraged me at just the right moments, I embarked upon this biography.

When it came to gathering facts, no one surpassed Tom Kelley in helping me reconstruct the early *Atlantis* cruises. Val Worthington is another fine source of anecdotes and information, and I am still amazed when I think of the generosity of Charlie Wheeler and Dana Densmore, who offered me their personal journals without a moment's hesitation or a hint of a restriction. For the improvement and correction of my manuscript I owe much to George and Marion Clarke, Mary Sears, and Al Vine, my most conscientious critics.

Much of the information in this book came from the Woods Hole Oceanographic Institution's data center and adjoining archives, where the letters, radiograms, charts, and photographs for each cruise are now stored. Bill Dunkle watches over these files, and during the past two years he has helped me by tracking down missing letters and forgotten facts, identifying people in photographs, and suggesting new sources of information.

Doris Haight did all the typing, Dave Owen printed most of the photos, and Sally Mavor did half the sketches that appear at the ends of the chapters. Bob Ruiter helped with photography. Art Maxwell smoothed out problems that arose between me and the Institution, and my husband, John, smoothed out a good many more. Jean Baur Walling, my sister, tried out her new editorial skills on the manuscript, and everyone who reads the book can be thankful for that.

If a large portion of the ship's story has come from old records, an even larger portion comes from the scientists, officers, and crew who sailed aboard *Atlantis*. To all with whom I sat down and talked — in their living rooms, laboratories, boathouses, offices, and hospitals — thank you.

Permission to quote from unpublished journals, letters, and manuscripts has been received from the Woods Hole Oceanographic Institution, Eleanor Iselin, Mary Bigelow Soutter, A. E. Parr, Harriet Ewing Doey, Allyn Vine, B. King Couper, Margaret Backus, Charlie Wheeler, Gilbert Oakley, Fred Phleger, Adrian Lane, Dean Bumpus, Eugene Mysona, Mary Bray Blackburn, John Pike, Dana Densmore, the Marine Biological Laboratory, Thomas N. Kelley, and the *Washington Post*.

Susan Schlee

Falmouth, Massachusetts

ON ALMOST ANY WIND

Introduction

In the summer of 1931 the steel-hulled ketch *Atlantis* became the first deep-water research vessel for the new Woods Hole Oceanographic Institution on Cape Cod, and the serious investigation of the North Atlantic Ocean, which had come to a standstill some fifty years before, resumed. There had been a time in the nineteenth century when the study of the deep sea seemed to promise great rewards and several large oceanographic vessels had been engaged in research, but by the 1880s interest had declined, and by 1920 the last of what might be called first-generation research vessels had been scrapped or retired. Oceanographers were left without a means of studying the open ocean.

With the arrival of *Atlantis* a new round of investigations began, and during the next thirty-five years the ketch made 299 cruises over more than a million and a half miles of ocean. Because for six years she was the only vessel in the country large enough to undertake such extensive investigations, and because for another fifteen years she was one of only a few such ships, it was inevitable that scores of important discoveries were made from her decks. In fact, during her exceptionally long career, from 1931 to 1966, the ketch all but monopolized the testing of new instruments and the exploration of uncharted regions.

Beginning with a stormy maiden voyage on which cases of beer jammed the ship's steering mechanism, the mate treated three serious injuries, experimental nets fouled and tangled, sewage tanks backed up, the stove broke down, and friendships were made that are still unbroken, the story of *Atlantis* covers an incredible variety of cruises. There were short test cruises on which gun corers, deep-sea trawls, and plankton samplers were tried, and there were long ones on which the Caribbean, the equatorial Atlantic, and the Mediterranean were surveyed. There were cold trips to the fogbound banks off Nova Scotia and hot ones to the glaring tropics.

In the 1930s scientists at the Institution pursued a program of basic oceanographic research, and for almost every study the new *Atlantis* was essential. From her decks geologists sounded deep-sea canyons and took the measure of massive undersea mountain ranges; physical oceanographers sampled the Gulf Stream; and biologists brought up animals never seen before. Later in the thirties, applied research related to submarine warfare was added to the basic studies, and as World War II approached, *Atlantis* worked with destroyers and submarines to perfect instruments and techniques that eventually saved the lives of many American sailors. When the Germans brought the war across the Atlantic, the ketch was forced to retire from military work, for with her mainmast rising 140 feet and her full keel extending 18 feet below the waterline, she was an easy target for the German submarines cruising off the eastern seaboard. Consequently, *Atlantis* was quietly sent out of Woods Hole to sail without lights to safer waters. Six lookouts were posted each night as the ship moved through the darkness, yet just off Cape Hatteras a Standard Oil tanker emerged from a thick haze and the two vessels collided in a shower of sparks. By the time the ketch had been repaired and sent into the Gulf of Mexico, the Germans were there too, forcing her to retreat to Lake Charles, Louisiana, where she was laid up for the rest of the war.

When the ship reemerged in 1945, oceanography had changed tremendously. Having come to the attention of the navy, and having proved useful during the war, the young science abruptly received the funds it had always lacked. Oceanographic facilities doubled, then tripled, and oceanography itself evolved from a small cluster of disciplines centered around marine biology to a "big science" concerned in large part with the physical processes of the sea and their effects on underwater warfare. Like other such laboratories, the Woods Hole Oceanographic Institution expanded dramatically during the war, but although it acquired a veritable fleet of smaller vessels, *Atlantis* was still the only ship capable of offshore work. The ketch was therefore sent out on cruises as fast as she could be gotten ready, and instead of working only in the western North Atlantic, she was sent to the Mediterranean, the South Atlantic, and finally into the Pacific.

Under the energetic direction of Maurice Ewing, one of the most famous geophysicists of this century, *Atlantis* made a series of voyages to the Mid-Atlantic Ridge and brought back the first precise bathymetric charts and samples of rock. Between two of Ewing's cruises the ketch made her longest single voyage, to the Mediterranean and Aegean seas, ostensibly to study local fisheries but actually to make classified bathymetric charts for the navy.

In the early 1950s cruises were launched to explore the trade winds belt, and although *Atlantis* was being used more and more as a powered vessel and less as a sailing ship, there were memorable nights in the tropics when her wrinkled sails were set and, with her lee rail under, she drove through a phosphorescent sea. Later still, in 1958 and '59, a multinational exploration of the sea was launched during the International Geophysical Year. As her part in this program *Atlantis* made extensive voyages to the Mediterranean, the Red Sea, and the South Atlantic. The Red Sea cruise was a particularly difficult one. The heat was so intense that most of the men got sick, a fire broke out on board within spitting distance of several tons of explosives, the captain was burned, and the chief scientist, bitten by a dog, was given fourteen rabies shots by a man who, it was rumored, disliked him so much that he tried to make the shot marks read "SOB."

But even as the ship's territory expanded and her work increased, it became obvious that an aging sailing vessel was neither large nor powerful enough to take on the newer and more demanding oceanographic investigations that had developed during the war. By the late 1950s, large, fully powered research vessels were moving into Woods Hole, and with the advent of *Chain* and *Atlantis II* the ketch became "the old *Atlantis*" both in name and in fact. Still, there was such sentiment for this ship, as a symbol both of the early, adventurous days of oceanography and of a more intimate relationship between man and sea, that some scientists at Woods Hole went out of their way to keep the ketch in business. By 1962, however, *Atlantis* was so obviously antiquated that government contracts were sometimes awarded the Institution with the stipulation that the old *Atlantis* not be used. Nevertheless, she found a temporary reprieve from retirement as a training ship, and for two summers carried students through the waters around Cape Cod.

In 1966 *Atlantis* was finally sold. After completing her 299th cruise — having sailed across one and a half million miles of ocean — the vessel was repaired, repainted, and rechristened *El Austral*. Before sailing to Argentina, where she would again work as a research vessel, she made a last visit to Woods Hole. It was a gray November day when, after a brief ceremony, the vessel moved for the last time through the slicks and eddies of Great Harbor. The people of Woods Hole watched with sadness. A beautiful sailing ship was leaving, an exciting period of oceanography was coming to an end.

On Almost Any Wind is first of all a sea story. Based on the logs, letters, and memories of the ship's company, it is an account of the adventures of America's most renowned research vessel. Second, the book is a biography, an attempt to follow the life of a ship as it developed in response to the fortunes of oceanography and as it exerted its own strong influence on the men who lived and worked aboard her. Finally, it is a mirror in which parts of the history of oceanography are reflected.

The construction of *Atlantis* marked a renewed interest in marine science. In 1927 the National Academy of Science undertook a study of the nation's oceanographic needs and resources and found that although there were three or four government bureaus involved and a half-dozen private or partially state-supported institutions — such as the Scripps Institution of Oceanography, which was certainly the center of American oceanography at the time, although it owned only a converted purse seiner for coastal work — there was no facility for deep-sea research on the east coast of the United States. To correct this imbalance, the Woods Hole Oceanographic Institution was incorporated in 1930 and, as its founders had been led to expect, given approximately $2.5 million by the Rockefeller Foundation. A laboratory building was soon going up at the edge of Great Harbor, and to the envious surprise of scientists at other institutions, a new research vessel was being designed and built.

The Institution's first director, Henry Bryant Bigelow, hired the prestigious firm of Owen and Minot to design the vessel, but it was Bigelow, not the firm, that decided it should be a sailing ship. Although by 1930 oceanographers recognized the need for a ship that could take them swiftly across the sea through almost any weather and had the power to handle heavy equipment, Bigelow realized that it would be close to impossible for a small, private institution to build and operate such a powerful vessel. Supporting this financial argument in favor of a sailing ship were the added considerations of a sailing vessel's greater range (it was not dependent on fuel) and greater stability. This last consideration would allow scientists to work at their microscopes or with chemical apparatus even in rough seas.

A less obvious reason for the choice of a sailing vessel was that both

Bigelow and his twenty-six-year-old student Columbus O'Donnell Iselin, already chosen as the first master of *Atlantis*, had an aristocratic preference for sailing. Both men came from well-connected families and had sailed all their lives. Both were part of the elite Harvard community with its tradition of pursuing knowledge — even in the ocean depths — with elegance and style. Bigelow had used motorized research vessels, which by their very nature could largely disregard the temper of the sea. He had also worked at length from a schooner and knew well that to study the ocean from a ship that had to adjust continuously to every aspect of the ocean was to force the researcher to adapt his schedule and technique to the very thing he was studying.

When plans were completed for what Bigelow called "a sweet ship," bids were let on her construction, and because it cost far less to build abroad, the contract was awarded to Burmeister & Wain in Denmark.

"We have just had a wire from Copenhagen that the ship's keel has been laid," wrote Bigelow in October 1930, and on the last day in December the ship was christened *Atlantis* and launched.

The entire shipbuilding community at Burmeister & Wain gathered for the launching, for Marconi-rigged ships with their towering masts were a novelty, and many builders had never seen, much less built, a ship with masts 140 and 100 feet high. Some believed the vessel would roll over and sink as soon as those masts were stepped, or if not then, as soon as the first North Atlantic gale laid the ship on her side. The launching, at any rate, was a success, and in March of 1931 Bigelow visited the ship and proclaimed her "a *beauty!*"

Atlantis was a double-ended steel-hulled ketch, the largest such vessel in the world. She had a length overall of 142 feet, a beam of 29 feet, and an original draft of 17; she displaced 460 tons. Rigged as a ketch, she could carry 7,200 square feet of canvas — an amount modest for her hull size but ideally adapted to heaving to. Even more modest was her 280-horsepower diesel engine, which could drive the ship at seven or eight knots and was also used to power a heavy trawl winch carrying some five miles of half-inch steel cable. A smaller diesel provided power for light, refrigeration, ventilation, and a hydrographic winch.

Late in the spring of 1931 the sea trials of *Atlantis* were begun in the sounds near Copenhagen, and Columbus Iselin went to Denmark to accept the vessel for the Oceanographic Institution. Iselin ("one of the yachting Iselins," some people liked to point out, thinking mainly of his granduncle, who had four times defended the *America*'s cup) was a handsome sight as he stood at the wheel of the new research vessel in a

Atlantis under construction at the Burmeister & Wain yard in Copenhagen. *(Courtesy Woods Hole Oceanographic Institution.)*

smart, English-made uniform. The first mate, who was later accused of confusing solemnity with pomposity, wore a sword for the occasion.

With the ship formally accepted, provisions for her maiden voyage were carried aboard. There were chemicals, laboratory glassware, and plankton nets; linens, pots, pans, and dishes; potatoes, meats, milk, and case after case of Carlsberg beer.

The young crew was assembling, too — college friends of Iselin's who had sailed with him before, professional sailors from Denmark, Norway, and Finland, a short-order cook from Hudson, Massachusetts, an engineer from England, Henry Bigelow's brother and eldest son, and Iselin's cousin by marriage, the charming and irrepressible Terrence Keogh.

With the stores loaded and a good part of the crew aboard, *Atlantis* sailed from Copenhagen to Plymouth, on England's southwest coast, where she picked up the remainder of her crew and scientific staff. It was the middle of July 1931, and her maiden voyage was about to begin.

1 Maiden Voyage

So I began the study of the sea. . . . And this work has filled the most wonderful years of my life. Today, I regret no effort I have spent on it, for its influence has kept me from pain and folly and has mitigated the sorrow that little by little fills that portion of the heart which originally abounds with dreams of happiness.

— Albert I, prince of Monaco

July 16, 1931, was a Thursday, and in Plymouth not a particularly pleasant one. A light rain fell from clouds that hung low over the harbor, and as a small knot of people gathered halfway up a hill, a launch transferred several men and their luggage to a graceful white ketch anchored far from shore. Immediately a group of young men on deck began to pump a seesaw-like windlass, and with a long rattling complaint, the ship's anchor was drawn up from the bottom of the bay. The small figure of a pilot jumped agilely to the chartroom roof, and in accordance with inaudible directions the jib and mizzen were sheeted in. The vessel heeled gracefully to the west wind and in a moment was sliding along the shore. She was headed for the English Channel, for the open Atlantic, for Boston, and ultimately for her home port on Cape Cod, where she would serve the Woods Hole Oceanographic Institution. The maiden voyage of the research vessel *Atlantis* had begun.

Shortly after 4:00 P.M. on what had been for many an anxious and inauspicious day, *Atlantis* rounded the Plymouth breakwater. Her second headsail went flapping up the forestay and the vessel began beating to the southwest. Along her decks sailors carried luggage and crates of scientific equipment while in the deck laboratory amidships and in the larger laboratory below, chemicals and glassware were secured alongside piles of lumber that would eventually be made into cabinets.

For Columbus Iselin, the ship's twenty-six-year-old master, the first several days out of Plymouth were misery. The west wind increased all

20

that first afternoon, and by evening, when *Atlantis* joined the traffic in the English Channel, she was dancing uncomfortably over a choppy sea and throwing water back along her decks like a vessel half her size. Iselin was not sure how the ketch would behave in foul weather or how well his deck officers could handle her, and he was reluctant to retire. He finally fell asleep at 4:00 A.M.

When Iselin clambered back on deck later the same morning, he found to his dismay that the ketch was sailing on her ear. Icy water ran foaming and swirling along her lee deck and the ketch was slogging ahead with a disagreeable pitch and roll. Predictably, most of the ship's company had gone below and were wedged into pipe berths less than two feet wide. They were feeling simply awful.

By the second afternoon the wind had risen to gale force and the seas were mountainous. Having cleared the Cornish coast, *Atlantis* was, in theory at least, heading northwest toward a point off Iceland where the scientific work would begin. In fact, she was going almost nowhere. With her foresails and mizzen still set, she was pinned on her side by a booming wind that snatched the tops off the gray waves and threw foam in streaks along the water. To windward, sixteen- and eighteen-foot waves rose glistening like black hills before the ship. Hissing and curling, they advanced to confront the ship's prow, hung menacingly above her, then with some last-minute trickery disappeared beneath the vessel. Down into the next trough, up onto the next wave, and all the while she lay scuppers under with torrents of water, as Second Officer Joseph Bigelow had just discovered, somehow pouring into the holding tank and backing up through every basin, tub, and toilet on the ship. Water was also leaking into the chartroom around the wheelhouse windows.

Ordering the holding tank pumped continuously, a nasty job that involved going down into the bilge and sweating up and down over a fitting that did not seal off the odors, Iselin scrambled down into the flooded lower laboratory and, with mop and bucket, began making passes at the regurgitated sewage that rolled and bumped across the lower deck as the ship heaved. As he slid back and forth, he wondered how many other systems on *Atlantis* had been installed as if the ketch were designed to sail upright like a steamship.

After tracing pipes all over the ship, Iselin and his officers found two small vent pipes leading from the sewage tanks to the ship's side. First Officer Samuel Clowser and Engineer Harold Backus cut and plugged while Iselin pumped and bailed. Still salt water entered the tank. More searching revealed a third vent. When this too was

When *Atlantis* sailed from Plymouth, England, in 1931, she encountered the same weather that had greeted *Beagle* when Charles Darwin set sail from the same port exactly 100 years earlier. "If it was not for seasickness the whole world would be sailors," he wrote. *(Courtesy Woods Hole Oceanographic Institution.)*

22

plugged, the pumping went down to about seventy-five strokes a watch and the washrooms became usable.

The crew did not recover so easily. Iselin helped one of the cooks to prepare some supper, but fewer than half the men were interested in eating, and fewer than that after they watched their food advance and retreat with the undulating motion of the gimbeled table.

After supper, seasick and well alike were called out on the stormy deck to bring *Atlantis* about on a new tack. Men were needed to handle the sheets, and since storage cabinets had not yet been built in the laboratories, others had to push the glassware uphill from the lee side to the windward side so that when the ship came about, the glass would again be on the low side. Once these stores were shifted, the word was given and the helm put hard alee. The ship came into the wind, and as she straightened, her booms swung in and her sails could be heard banging and snapping above the noise of the storm. But she would not fall off on the new tack. With a lurch the ship resumed her original course, and in the laboratories the glassware skated across the slippery decks and crashed into the bulkheads. Again the order was given to come about and again there was a great flapping and banging, but the ship refused to tack. The trouble seemed to be in the steering mechanism: the wheel would not turn past a certain point. Clinging to rigging, rails, and bulkheads, Iselin made his way back to the pilot-house and from the chartroom just forward went down into the compartment where worm gears connected wheel to rudder. The space on either side of the steering mechanism had been used to store stacks of beer cases, and a glance showed that one stack had tilted into the gears. These cases were hurriedly shoved aside, and on the third try *Atlantis* came about and filled away on the new tack with such speed and force that ten men could only just manage to trim the jib.

"We are making slow progress," Iselin glumly reported in his journal, "and of course not nearly in the right direction. All in all it has been a pretty blue start."

Although *Atlantis* had just begun her maiden voyage, most of the twenty-five men aboard had already become fairly well acquainted, more by sharing the inevitable follies ashore than by working together at sea. Eighteen or nineteen of them had gathered earlier that summer in Copenhagen, where the steel-hulled ketch had been built, and while the vessel was undergoing sea trials, had lived in a hotel doing what little they could on the ship and spending most of their time sightseeing.

Columbus O'Donnell Iselin, first master of *Atlantis*. *(Courtesy Woods Hole Oceanographic Institution.)*

The ship's company was divided into three groups — officers, seamen, and scientists; or, perceived from a different angle, permanent employees, visitors, and friends. The Institution's director, Henry Bigelow, had wanted to hire a nucleus of professional men who would stay on the ship for several years, add several young scientists similarly committed to the Institution, and fill out the remainder of the ship's company with friends and relations. Many in the last category had sailed with Iselin before on the adventurous scientific expeditions he had organized on his own schooners. In accordance with Bigelow's plan, three professional Scandinavian sailors had been hired, and more than all the others combined, these three ran the ship and provided models of excellent seamanship. In the engineroom the man soon in charge was a Yorkshireman, Harold Backus, and among the deck officers only First Officer Samuel Clowser was expected to stay on.

Clowser, another Britisher, had been lured away from a secure position with the Cunard Line, and it was intended that he soon become master of *Atlantis*, thus allowing Iselin to spend more time on science. But having received a British seaman's strict and well-disciplined training, Clowser had no sympathy for the irreverent informality that Iselin's friends brought with them; and with what many described as overbearing pride and unnatural persistence, he tried to instill the Cunard Line's etiquette in the crew. Clowser insisted that the crew remain before the mast (as in Richard Henry Dana's time), and he ordered communication between scientists and crew to pass with formality through the officer on watch. He prefaced so many orders with "The way we did it on the Cunards . . ." that the phrase became a standing joke among the boisterous crew.

One of the Scandinavian sailors had reacted by buying an armload of old naval uniforms from a secondhand shop in Plymouth, and as Clowser smartly stood the 8:00-to-12:00 or captain's watch, the sailors emerged from the fo'c'sle dressed in knickers, coats, and immense fore-and-aft hats, and paraded with banners before the mast. (To get their money's worth, they used the costumes again when they hired an open horse-drawn coach and rode, nodding and waving benignly, through the crowd that strolled along Plymouth's waterfront.)

But the able seamen did not put the wit and cunning into battling Clowser that the brash Irish bosun, Terrence Keogh, did. Terry was Iselin's cousin through marriage, and having grown up with him in the same seaside town, had sailed with him for many years. Both were fine deep-water sailors, but here the similarity ended, for Iselin was as steady and responsible as Terry was bursting with irrepressible energies. Apparently Terry's father, Judge Martin Jerome Keogh, a judge on the Superior Court of New York State and a teetotaler, had made a strong and entirely negative impression upon his son. Terry had little respect for any law and he drank an outrageous amount at every opportunity. Iselin knew this, of course, and was not surprised that by the time *Atlantis* had arrived in Plymouth, Terry had already stirred up a great deal of trouble. He had been arrested for drunk and disorderly conduct in Copenhagen (and pulled through the skylight of the jail just a few hours before sailing); had nicknamed Clowser "bold Physic Face," to the delight of the crew; and just as he had once expressed his good spirits by throwing firecrackers and railroad flares into bunks on Iselin's schooner *Chance*, he had heaved one of Clowser's several pairs of long underwear off the stern of the ship. He had, in short, caused so much hilarity and consternation that upon arrival in Plymouth Iselin tried hard to keep him from going ashore.

For a while he had been successful, but inevitably Terry wore away his resolve with artful flattery and implausible promises, and a few days before sailing time Iselin had allowed him to go ashore to pick up equipment. He disappeared immediately. Terry visited a number of pubs, it seems — charmed the barmen, the patrons, whoever happened to listen to his stories — and when night fell climbed aboard a bus to Plimpton to get some sleep. Terry made the round trip between Plymouth and Plimpton until the bus service ended for the night and he was turned out in the center of Plimpton. He staggered into a public lavatory and, finding the acoustics to his liking, began singing in his fine rich voice with its cultivated accent, first the silliest, then the bawdiest ballads he knew.

> And I says, "O Captain mine" —
> My eyes were runnin' brine —
> "Your evil thoughts of me
>   Give me sorrow.
> "I'm going ashore," says I,
> "I'll behave most proper-lie,
> And be with you at the dawnin'
>   Of the morrow."

Predictably, a bobby arrived, but Terry had locked himself into the stall and would not come out. The bobby angrily broke down the door, and just as he burst in, Terry wrenched the seat off the toilet and popped it over the policeman's head.

The following day a Royal Navy gig stood by *Atlantis* and in great style delivered Terry and the charge against him — "disturbing the peace and destroying His Majesty's privy." (It was later learned that

Terry had so delighted the police chief with his stories that he had not spent the night in jail, but in the chief's own home as his personal guest.)

On July 18 the west wind that had shaken the crew down so mercilessly blew itself out, and although the ship still rolled and pitched in heavy seas, she was noticeably steadier. No additional sail was raised, and, as Iselin had hoped, most of the crew began to recover. Many had not eaten for a day and a half, and one, who went forward to see what was cooking for supper, found Crawford, the cook, in tears in the galley and the whole forward part of the ship filling with smoke. Crawford, his face and apron already blackened and his hat knocked off, was swearing at the oil burner in the stove, which refused to stay lit.

"At first we thought that the stove was not getting proper draft," wrote Iselin, who had quickly been called. "We fussed with the stovepipes for about an hour with no results. We lit the burner about a thousand times, only to have it go out and fill the inside of the ship with nasty smoke. We became covered with soot and grease ourselves and got nowhere. Finally I told Crawford to give us cold supper and then went aft to think and recover because I have seldom been so baffled or so mad."

Late that night the clouds that had hung over the ship since her departure began to break up, revealing patches of starry sky. As usual, Iselin stayed awake until 4:00 A.M. to be on hand if Pen Higginson, the officer with the least experience, needed him during the middle watch. The ship was finally making fine headway and almost in the desired direction. She jumped along at eight knots, heeling only moderately and pounding hardly at all.

"In short, I am in somewhat better spirits now," wrote Iselin just before dawn, "especially as I now think I understand what is wrong with the stove."

Sunday, July 19. We certainly have not understood what is wrong with the stove. We thought that the flow of oil was not even, but now, after a whole day's work, we think that the draft is insufficient. Therefore McLunin [one of the engineers] and I spent the whole middle watch tonight making a long stovepipe out of sheet galvanized iron. We were very successful as far as making a good pipe goes, but do not know yet how much good it will do the stove.

.   .   .

I only turned in for about an hour after my watch last night because I was so anxious to know how the stovepipe works.

On returning to the wardroom Iselin found to his surprise that Harold Backus, the engineer about to go on watch, was not waiting around for the stove but had already started cooking bacon and eggs over a blowtorch. A blowtorch was not part of the ship's equipment and the gasoline to fuel it was not allowed belowdecks, yet Backus had apparently produced both from under his bunk.

"I said nothing to Harold and wedged myself in the downwind corner of the seat in the wardroom so as to hold the frying pan above my knees while he applied the torch in an expert manner to the under-side of the pan. . . . After eating we got the galley stove going before he went on watch at eight."

With the stove and the sewerage system fixed and the ship gliding along to the northwest on a fresh breeze, Iselin could begin helping the scientists, George Clarke, Franz Zorell, and their two assistants, set up the equipment they would need in less than a week, when the scientific stations or stops began. Iselin spent a large part of his twenty-hour day building a rack for the water bottles in the deck laboratory, and the following day he was pleased to see that Zorell and his assistant had converted empty wooden beer cases into a place for their five hundred water-sample bottles and that Clowser had partly finished four or five brass pots for the plankton nets.

Iselin considered himself responsible for the scientific portion of the voyage as well as for the safety of the ship herself, a double role he had undertaken at least three times before on his own schooners. But on *Atlantis* both the science and the sailing were on a much larger scale, and Iselin was exhausting himself trying to provide the same attention to each detail that he had on his own vessels. He was particularly un-easy about leaving the sailing to anyone else. It was not that he be-lieved the ship unseaworthy, but he was bothered by the lack of sailing experience of his deck officers. Even a month out of Plymouth he felt none had acquired any real "feel" for sailing.

"As a result I spend more time awake than I should, probably unnec-essarily. I hardly like them even to tack unless I am on deck. Nearly always something goes wrong. The sails are hardly ever trimmed right and are not handled smartly. The crew are willing enough, but nobody gives them the proper orders."

Such thoughts were in Iselin's mind that evening when the wind began to rise again and he could alternately hear the nervous rattle of luffing sails and feel the ship lengthen her stride of pitch and roll as she was let off the wind.

"About dark the wind picked up in fine shape and we were soon tearing along at ten or eleven knots. But it got rough almost imme-diately and the ship began to dance around like a small boat and throw water all over the place. Finally about eleven we struck a hard squall

The mainsail coming down. *(Dana Densmore photo.)*

and the mainsail had to be taken off. It must have been blowing forty miles an hour when we called out the whole crew and began to lower."

Rain spat down upon the slippery deck like hailstones, and even with floodlights on it was difficult to see and almost impossible to hear above the pounding and creaking of the ship and the racketing of the wind. Clowser was the officer on duty, but his orders were either not being given or not being heard, and Iselin watched in dismay as the ship caught a sudden gust and rolled far to leeward. The crest of a wave swept over the bow and the water ran swirling and foaming along the lee decks.

"Slack off the mainsheet!" Iselin shouted, and as the sailors still went sliding around the deck from handhold to handhold without apparent purpose, he moved to do it himself.

"As the sheet was pretty well off, while the men were at the lifts forward, I went to lead the sheet back to the hydrographic winch. I put a stopper on it and cast the sheet off the bollard. Thinking that the stopper was holding, I was just moving aft to the winch when *bang*, the sheet ran out like a great snake, taking a turn around my left leg and throwing me down and dragging me along the deck."

Iselin lay stunned for an instant, his leg crushed against the bollard. Above him the mainsail shook violently and the careening boom threatened to carry away. Clowser, speechless, watched as Iselin slowly began to crawl aft across the gleaming deck. The sailors hung desperately to the sail, and out of the darkness burst a torrent of rich profanities, followed by orders. Terry had the helm eased, and as *Atlantis* came up into the wind the mainsheet was brought under control and the heavy, sodden mainsail was lowered and dumped on deck.

"At first I thought my leg was broken," recalled Iselin. "It was very painful, but I did not lose a great deal of skin, thanks to my pants. The worst place seems to be my ankle bone, which got smashed against the bollard. Mr. Bigelow gave me a drink and I could just stand the iodine he put on. There is a great spiral groove around my leg where the rope squashed it. I think it is nothing worse than a very bad bruise and am really lucky not to have received a serious injury. It blew great guns all night with fierce rain squalls."

Lying sleepless in his pitching bunk, Iselin discovered that his cabin leaked.

Whatever may have crossed his mind that miserable night, it was probably not self-pity. He had been in worse situations at sea many times before. He had been on the water, in command of his own vessels, under all imaginable circumstances, for at least fifteen of his

After his accident, Iselin hobbled gingerly along the deck. Terry Keogh stands beside him. *(George L. Clarke photo.)*

twenty-six years. He had grown up in New York City and New Rochelle (on Long Island Sound), and at the age of eleven had built his first leaky boat, *Sponge.* He had sailed and rowed his way through St. Mark's preparatory school, and as an undergraduate at Harvard University had bought a captured rum-running schooner, *Theresa White.* In the summer of 1925, he and some college friends made a trip in her to Nova Scotia, and the following summer, after graduation, he took possession of the first boat built for him, the seventy-seven-foot schooner *Chance.* With Terry Keogh, Bart Hayes, and other friends, Iselin made an adventurous 5,000-mile voyage to the northern portion of Labrador to measure the temperature and salinity of the water and make other relatively simple oceanographic observations. He had been encouraged to do so by his friend and teacher Henry Bigelow, who four years later became the first director of the Woods Hole Oceanographic Institution.

31

In the summer of 1927, Iselin, then enrolled in a master's degree program at Harvard, had made a more scientifically sophisticated expedition on *Chance* to study the Gulf Stream. He was having a still larger schooner built, and in 1928 sailed to Europe and back aboard his new *Atlantis,* bringing back water samples, plankton, temperature data, and, in spite of Prohibition, a barrel of port wine for a friend who was getting married. Iselin himself was married the following year, and in 1930 Bigelow asked him to join the Woods Hole Oceanographic Institution as master of the research vessel *Atlantis* and, more permanently, as a physical oceanographer. (Iselin, who had sold his schooner *Atlantis* to Alexander Forbes, a trustee of the Institution, asked Forbes if he would rechristen the schooner so that her name could be transferred. The schooner thus became *Ramah* and the Institution's ship *Atlantis.*)

Iselin was ordinarily a good-looking man — he had dark curly hair, a fine build, and a shy smile — but on the day following his accident his unshaven face was an unhealthy white and his eyes were sunken. He looked extremely uncomfortable and extremely short of sleep.

"I stayed in my bunk all day," began the next several entries in his journal. A list of complaints followed. The headwinds continued. The stove broke down. The sewerage system failed again. "These are really very boring days. We get nowhere, there is no sun, it is cold and wet, and the boat jumps around too much to do any real work."

Iselin's leg throbbed and ached, and it was difficult for him to sleep. When he woke in the dark, as he often did, and heard the ship's bell ring twice or three times, he would switch on the light above his bunk and read for several hours. By eight bells, 4:00 A.M., when he could hear the men changing watch up on deck, he usually put down his book and tried to sleep again at least until dawn. During one such interrupted night he dreamed that the stays and shrouds holding the masts in place were slowly slipping through the great chain plates that held them to the ship's sides, for he believed their ends to be seized instead of properly spliced into permanent loops. Several mornings later, as soon as he could walk, Iselin appeared on deck with a can of paint and painted marks on each of the cables so that he could tell if in fact they were beginning to slip.

Five cold and rainy days after Iselin's accident the scientific work began. It was divided roughly into four parts. There were hydrographic studies concerned with the temperature and chemical properties of the water, light-intensity studies made to discover how far light penetrated into the sea (and how this depth of penetration affected the daily vertical migration of plankton), biological collections of plankton including eel larvae, and meteorological studies. Franz Zorell, a German who, Iselin suspected, had come along with the idea of showing the Americans how it was done, was in charge of the hydrographic work and was assisted by Bart Hayes. The two men wanted repeatedly to send down a string of water bottles, with thermometers attached, in order to record the temperature and procure a water sample from each of several intermediate layers in the ocean. This procedure would help them understand how masses of water of varying temperature, salinity, and oxygen content were arranged within the North Atlantic basin and would suggest how the waters circulated. Zorell had already mixed the chemicals he would need to test the water samples and, with an inimitable conglomeration of German, proper English, and profanity (the last taught him by Terry under the guise of "English lessons"), he ordered the hydrographic boom rigged and tested.

At 5:50 A.M. on July 26 the jib was lowered and the ship hove to. A brisk northeast wind made the men on deck shiver and slap their arms, but the sky was clear, and there was bright sunlight for the first time since they had left Plymouth.

The early temperature station went very well [Iselin reported] and we were finished shortly after breakfast. We then headed south under mizzen and jibs until about 11:30, when we stopped again to try the light-intensity and plankton gear. At once things began to go wrong.... In the first place, the clamps do not exactly fit our wire, so the two nets we put on were not free to turn around the cable. In the second place, it was impossible to get the head of the dredging boom [which was swung out at right angles to the ship when in use] far enough aft [inboard] to put the nets on the wire. We tried hauling it in with a tackle but had a struggle that finally ended in losing one net.

During the course of this operation the boom collapsed inboard and we had our second bad accident. Bart [who was leaning over the rail to fasten a net onto the cable that led from the boom] was caught between the boom and the rail and badly crushed. I'm afraid he may have broken a rib. It was not pleasant. There is no further use trying to place instruments on the wire while standing on deck. This was more fully confirmed in the evening when we tried the eel nets after finishing our second temperature station. We fought with the nets, the wire, the boom, and each other until twelve o'clock and only completed three hauls. There were no eels in any of the nets....

All this time I was forced to take an inactive part because of my leg. I sat on whichever cabin house was closest to the seat of struggle then in progress and very ineffectively tried to give advice. I am sure that if I had been on my feet I could have made things work.

On July 28, two days later, clouds again hung over *Atlantis* as she moved steadily southward on a course that would lead her straight down the Atlantic to the latitude of Bermuda. It was still cold and raw on deck, and the crew not on watch either sat around their mess table reading and talking or slept in the fo'c'sle. Aft, Zorell worked up the water samples taken during the morning's hydrographic station, and Iselin and George Clarke, a young biologist who had just received his Ph.D. from Harvard and who had sailed with Iselin aboard *Chance*, made plans for the evening eel tow. This project was part of the work begun by Johannes Schmidt, a famous Danish biologist, who had worked from 1904 until 1922 to delineate the spawning grounds of the European and American eels. He was still adding information to his fine studies and had asked Henry Bigelow if the men on *Atlantis*'s "virgin cruise," as he put it, could try to catch the elusive larvae.

Eel larvae are transparent creatures roughly the shape of a willow leaf but usually a great deal smaller. They must be fished for at night, for by the light of day they seem able to avoid slow-moving plankton nets. Thus in the evening Iselin had a latticework cover taken from a hatch and rigged into a platform that swung somewhat unsteadily outboard of the ship's rail and within reach of the dredging boom. After dark, the ship was hove to and a 1,500-pound weight was attached to the cable that led from the boom. When the weight was safely over the side, *Atlantis* started sailing again, and Clarke climbed out on the staging and began to attach a series of coarse, burlap-like nets, each about five feet in diameter and eight feet long.

"*Leptocephalus!*" sang out a stocky Finn who was absolutely fascinated by the Latin name for eels and who was lounging against the rail watching the proceedings. "Tonight ve catch *Leptocephalus*."

As Clarke attached each net to the wire, Iselin, now well enough to operate the main winch, let a predetermined length of wire grind off the great spool that turned in the lower hold. When the last of the four nets was sent down to the desired depth, the whole string was allowed to drift with the ship for about two hours. Iselin then hobbled back to his perch on top of the deck laboratory, from which he controlled the winch, and Clarke, aided by a giant Norwegian sailor, climbed out onto the lattice platform that rocked and dipped with the ship. As each net broke the surface of the dark water with a gentle splash and rose to the level of the platform, Clarke grabbed the pennant that connected the net to the clamp on the wire and led it inboard until a sailor at the rail could hook it with a boathook and pull it partially over the rail. The sides of the net were sluiced down with a hose and the cod end, with

George Clarke, right, stands on the platform near the dredging boom waiting for a closing net to be brought aboard. *(Jan Hahn photo.)*

the catch, was brought aboard. Only when this operation was safely accomplished did the Norwegian start swinging his forty-pound maul at the releasing pin that would free the net's clamp from the wire. When the ship rolled and he missed the pin, he lurched toward the end of the platform and Clarke had to grab him by the shirt to keep him aboard; when he timed his blow well and hit the pin, Clarke had to grab that before it flew out and dropped into the sea. It was a curious ballet, lit eerily by deck lights and by the beam of a long flashlight that Clarke tried somewhat desperately to hold under one arm.

The cod end of each eel net was doubled over and tied with marline, and as the line was undone, the luminescent catch came sloshing out into a wooden tub "like a horse shitting Christmas-tree ornaments," Terry Keogh observed from his perch on the laboratory trunk. Although to most of the sailors it was a nondescript mess, to the scientists the catch was recognizable as jellyfish, salps, copepods, a few fish, and perhaps but not perceptibly a few *Leptocephali*. The catch was preserved with Formalin, and the next morning, when there was light, the search for the eel larvae began.

Just aft of the deck laboratory on the starboard side was a relatively clear stretch of deck where the scientists, and those crewmen interested enough to slip past Clowser, sat around the tubs and with tweezers or the point of a pencil tried to fish out the larvae. On impulse, an eel club was founded, open to anyone who could find at least two specimens.

Wednesday, July 29, was one of those rare mornings when the sun shone and the sky was so clear that it seemed it could never be closed in by clouds again.

"I got up on deck and tried to dry up my leg, which refuses to form a good scab," wrote Iselin. "About two o'clock a steamer was sighted hull down on the eastern horizon. We made out soon that she was very large and going at great speed. She then changed her course so as to pass close to us. She proved to be *Bremen* [the 50,000-ton North German Lloyd liner that had broken transatlantic speed records by racing across at twenty-eight knots] and we got out our signal flags to ask to be reported."

All hands came up on deck as *Bremen* neared, black smoke pouring from her stacks. The liner approached so rapidly that it seemed to expand rather than move forward, and soon throngs of passengers could be seen crowding her rail.

"There was a good deal of waving and flag dipping as she flashed past . . . ," continued Iselin. "We were hove to for a temperature station

. . . with the trysail dumped on the deck forward and the jib half down. She may have thought us in distress, but I rather think she changed her course to give the passengers a thrill."

Since *Atlantis* had no radio, she used her signal flags to ask each large ship she encountered to report her position. Henry Bigelow had arranged to have a New York newspaper publish all such sightings, but only one report appeared. *Bremen* reported (with what Clowser said was typical precision) that *Atlantis* was "west of Ireland."

The promise of the clear morning did not hold good, and by the time *Bremen* had passed, the sky was clouding over and the wind was rising from the northwest. The next day brought more of the same and Iselin had to shorten sail to ease the ship's motion, even though the wind on the beam blew only twenty miles an hour.

"It is an extraordinary thing how the architects could have made such a powerful-looking boat so tender," he wrote with annoyance. "With the weather we have had so far there has been no use for the mainsail. Either it blows a good breeze and we do not need it or the wind is light and it is better to use power. In other words, in average North Atlantic weather, the boat is not capable of standing up easily under her present rig and has to be nursed along like a yacht." (The ship needed an additional thirty-seven tons of ballast, and when Iselin added to this consideration the numerous leaks and poorly designed stove and sewerage system, he surmised that "the drawing office forgot that they were building a sailing vessel.")

Iselin's problems as master — testing the sailing qualities of *Atlantis* and the sailing capabilities of the crew — were similar to those confronting him as chief scientist, and both kinds stemmed from the newness of the operation. Although Iselin had combined sailing and science before over portions of the western North Atlantic, he had been in small boats that he could sail almost single-handedly, with a crew composed of friends and with scientific objectives as informal as they were limited. The maiden voyage of *Atlantis* was sailing and science on a much larger scale. Iselin was having to learn to let others sail the ketch without him, and he was forced to settle quarrels that inevitably arose among twenty-five men confined in a small space. Most difficult of all, he was expected to raise the scientific program from an amateur level to a professional one. None of the transitions was easy. It is remarkable that he accomplished what he did.

Saturday, August 1, was the first really perfect day of the voyage. *Atlantis* had almost reached the latitude of southern France, and the sun that morning had real warmth to it and the air felt soft. After more than two weeks at sea a routine had finally evolved, and when Terry Keogh came up on the empty deck before breakfast and saw the sun well risen and felt the ship moving lazily along, it struck him that this was a good way to live.

After feeling the air and glancing over the rigging, as he did each morning, Terry jumped down through the forward companionway, stuck his head in the fo'c'sle, and woke the next watch with a string of imaginative profanities, punctuated by the smashing of a beer bottle on the iron stanchion at the entrance to the fo'c'sle.

Laughing, and sleepily trying to answer in kind, the sailors on the next watch swung down from their pipe berths and within ten or fifteen minutes were up on deck ready to start the cleaning, scraping, painting, varnishing, and repairing that had to be done whenever the weather permitted.

Although Terry was the bosun and had immediate charge of the men, Clowser decided what work had to be done and he apparently believed in the old whalers' commandment, "Six days shalt thou work and do all thou art able, and on the seventh, holystone the deck and clean and scrape the cable."

About an hour before breakfast the ship was hove to for the morning temperature station, and as the water bottles were being lowered to collect samples of seawater the smell of bacon and coffee came up through the new stovepipe and floated back along the deck. By seven-thirty the ship was once more sailing southward and the sailors due to come on watch at eight piled into the crew's mess so that they could eat before relieving those who had been up since four. In the scientists' and officers' mess abaft the galley, Iselin, Clarke, Clowser, and others wandered in at odd times during the breakfast hour, ordered their eggs (any way except poached — the eggs were too old to hold together), and sat talking and smoking at the gimbeled table.

During the morning, as the sun grew warmer and the wind dropped, the engine was started to keep the ship moving. Terry had the steps to the engineroom scraped and showed a sailor how to fix the poorly fitting clamps on the plankton nets. Another man began to wash down the decks and Terry was abruptly moved to give him such a shove that a stream of cold water shot into the open ventilator that led to Clowser's cabin. This leisurely schedule continued until shortly before noon, when the ship was again hove to, this time for Clarke's light-intensity measurements. While a watertight case containing photoelectric cells was let down from the flagpole at the stern of the ship, one of the sailors suggested a swim. Since none of the men had had a chance to bathe since leaving Plymouth, nearly all of them stripped off their clothes on the spot, briefly reflected that a mile or more of seawater would lie beneath them, and dived or cannonballed off the rail. A shout went up as the first bare bodies hit the water. Iselin, whose leg had still not formed enough of a scab to risk swimming, helped rig a ladder, and soon a continuous procession of swimmers was climbing up the ladder, balancing along the rail, and leaping into the water again, using every trick remembered from a dozen swimming holes. When the light-intensity measurements were completed, and it was found that the sun's irradiation was reduced to 1 percent at a depth of about 150 feet, the swimmers came back on board. Dripping along the wooden deck, they gathered their clothes and went below.

After dinner at noon, the main meal of the day, the men who had been up part of the night piled into their narrow bunks in the fo'c'sle, where the sound of water rushing past their heads and the occasional thumps of unhurried footfalls on the deck above soon put them to sleep. Others went on watch, and a few came back up on deck either to wash their clothes in a bucket, wring them over the side, and hang them to dry on the rigging or to cut each other's hair or lie in the sun half asleep or simply to sit watching the shadows of the masts and rigging pass back and forth across the deck as the ship rolled gently on her way.

The afternoon passed quietly with another temperature station and swimming party late in the day, and at five-thirty Iselin, the officers, and several of the scientists gathered in one of the cabins aft to have a few drinks before supper.

"It is a very pleasant occasion," wrote Iselin, "and it gives us all a chance not only to tell dirty stories but also to discuss how things are going on board."

Supper was served at six, and several hours later, when the sun was down and the sky growing dark, *Atlantis* was hove to for an eel tow. On this particular evening the nets were sent down twice as deep as usual. The shallowest, towing at 100 meters, brought back three larvae, the deeper nets only a few mangled remnants of deep-sea fish.

By 11:30 P.M. the eel tow was finished, the deck cleared, and the ship on her way again. At midnight the helmsman rang eight bells and the watch changed. Lookout replaced lookout in the bow, helmsman replaced helmsman with a careful repetition of course heading, and Clowser, relieved by Higginson, left the chartroom where he had been plotting the ship's steady progress toward the Azores.

The sailors going off watch, like those coming on, were hungry, and as usual they were more than just annoyed that Clowser had proposed, with the cooks' consent, that no night lunch be left in the mess. A pot of coffee was usually there — in fact, it was the duty of one of the sailors going off watch to brew a fresh pot for the next watch — but there were no plates of cake, bread, cheese, or fruit left out. Consequently, as the ship moved through the darkness, two or three of the sailors quietly let themselves down into the bilges and helped themselves to cans of lobster meat and preserved fruits. Ordinarily they would have dropped through a skylight directly into the galley to see what they could find in the lockers there, but tonight the cooks had decided to do their baking when it was cool and the sailors could dimly see them as they passed in front of the range or stooped to look in the ovens.

Originally Henry Bigelow and Iselin had estimated that all the cooking could be done by a single cook aided by a messboy. Like so many of their financial short cuts — no doctor, no radio, no showers — the understaffing of the galley was based on what they had personally learned to put up with on much shorter cruises. Their plans soon changed. First, Mrs. Iselin's Scottish nanny threatened to quit if her brother Jock weren't brought over on *Atlantis* as a second messboy, and later a second cook had been added when it looked as though the first was going to be seasick a large portion of the time. Therefore a temporary bunk was built for Jock in a storage area containing hundreds of aging eggs, and a second cook moved in with the first in a cabin abaft the galley. In fact, the two locked themselves in for hours at a time, and as the voyage progressed it had become clear that the cooks were lovers. Some of the crew accepted this situation as a rather curious matter of fact; others felt it needed to be commented upon with shouts or blasts of water shot down the ventilator and with an occasional spying party sent to look through the galley skylight. Although the cooks did not rise to any of this baiting, the combination of their involvement with each other, their refusal to serve night lunch, and the progressive failure of the refrigerator created considerable tension forward. Each Sunday (when the crew did not work regardless of Clowser's beliefs) Iselin sent a bottle of whiskey to the fo'c'sle in the evening to soothe the crew's feelings, but "the cook came back to see me and I found that he had not been offered any." And so it went.

This routine, with its minor variations, continued as *Atlantis* rolled southward. Each day the temperature rose into the upper seventies, the wind blew gently, and the ship moved on toward the Azores. Clowser estimated that the islands should come into view on Monday,

August 3, and all that morning the sailors squinted along the horizon hoping to be the first to sight land. Not till early afternoon did someone point in astonishment several degrees above the horizon and there, floating above the haze, was the top of a mountain. By evening the ship was sailing between the islands of Corvo and Flores. A bank of clouds was piled up over the mountainous islands, and as *Atlantis* slipped close round the beautiful shore of Flores, rain could be seen high on the dark-green hills. Below, the setting sun still shone on a settlement of white houses surrounded by an irregular pattern of cultivated fields.

"Just before dark," Iselin wrote in his journal that night, "we sighted the dead body of a whale floating high out of the water and smelling higher for miles around. ('Who left the refrigerator door open?' yelled a sailor before he spotted the whale.) The evening eel-netting party was more successful than usual in that we got about fifteen eels and some new stuff, especially the larval stage of, I think, a conger eel."

Still within sight of the Azores, but with no intention of putting into a port, the men aboard *Atlantis* continued their routine, painting and varnishing parts of the ship and making hydrographic casts, eel tows, and light-intensity measurements.

Shortly after noon on August 4 Iselin was reading contentedly in his cabin when Clowser burst in with the news that Harold Backus, one of the engineers, had just fallen down the main hatch and was lying half conscious on the winch-room floor. Iselin ran forward and found Backus in great pain but with no apparent injuries. The stairs to the engineroom had been painted, and while they were wet the engineers had been dropping down through the main hatch into the room occupied by the wire reeling drum and going from there into the engineroom. Apparently Backus had leaned on the grating propped over part of the hatch and it had given under his weight, spilling him through the upper and lower holds. He had landed on his back on the winch below.

At first we made a sort of a bed of waste and sacks in the winch room [wrote Iselin], but he showed no signs of recovering and we decided to move him to his cabin. This was no easy job. We got a chair and put him on it. He made a great fuss and then fainted dead away. We had the six strongest men on the boat pass the chair up into the hold and then along to his cabin.

After thinking the whole situation over and watching him for about an hour, I felt that he had a concussion and that somehow, being out of his head, he thought himself back in the war, where he received a severe wound in his back. We put ice on his head and he fell into a heavy sleep. He is still sleeping and has only come to two or three times for an instant. He is no longer in pain,

Chief Engineer Harold Backus in his engineroom. *(David M. Owen photo. Courtesy Woods Hole Oceanographic Institution.)*

his pulse has remained normal, and his breathing is regular. There is nothing to do but wait until he comes to enough so we can make some kind of examination.

(It was not known then, but Harold Backus had fractured his skull. Years later, when he received a second blow on the head — this time from a hatch cover that blew shut in a gust of wind — X rays showed the lines of the old fracture.)

"In order not to increase our distance from the Azores and hospitals, we are making a haul for deep-sea fish with the fifteen-foot net tonight. We rigged the net in about an hour as all hands turned to to help. It went over the side easily and seems to tow without much drag."

The net, towed more than a half mile down all through the night, returned the next morning with only a few dozen fish.

Two days after his accident Backus seemed better, and although his back was very sore and he got dizzy whenever he tried to get up, Iselin decided that it was safe to leave the Azores and head west across the Sargasso Sea. Just before dark on the sixth, therefore, the mainsail was set and the ship headed west.

No sooner had we done this than the wind strengthened and we began to dash along at 10½ knots. By midnight spray began to fly around and a low bank of clouds covered the sky. About three o'clock we were caught by a heavy rain squall with every port and ventilator open. I thought for a while we would have to lower but we got through all right by running off for half an hour. During the excitement the galley got pretty well flooded and Crawford cut his stomach when he came on deck to take down the various rigs which he has to get fresh air into his place of business.

The days were growing steadily warmer now, and when the engine was run, the cabins, galley, and messrooms amidships became unbearably hot. It was 100° F. in the engineroom itself, and the engineers had prickly heat. At night the heat was less of a problem because there was often a breeze and *Atlantis*, running close-hauled, put her scuppers under and tore along at a fine rate.

For the next ten days *Atlantis* plugged along to the west, steaming through the still, hot days and beating to windward in light airs during the nights. Bits of sargasso weed floated past, and when the ship was on station swimmers kept a lookout for Portuguese men-of-war and sharks. Whole schools of flying fish appeared and, when startled by the ship, dashed off over the surface at great speed.

On the thirteenth the monotony was broken with a sudden crash. The engine stopped, and it was soon discovered that the governor shaft had parted, smashing a piston. The engineers worked for six hours "in a temperature of over 100°. We supplied them with beer during the day, but how they could work under such conditions of heat is beyond me."

On the night of the fifteenth a shark was sighted circling in the floodlight under the platform where Clarke was detaching the eel nets. One of the sailors baited a heavy hook and after a brief skirmish had the shark on board. It proved to be a female of a common species. Iselin cut her open, as she looked particularly thin, and found absolutely nothing in her stomach or gut. Evidently she had been very hungry. Later the long dried tail of the shark was lashed to the headstay. According to the Scandinavians, it would bring the vessel luck.

"The only other notable event of the day," Iselin wrote, "was that I shaved off my beard. This morning I got some graphite grease in it and suddenly, with great violence, I shaved the whole lot off. I must admit I am more comfortable now."

By August 18, Iselin had maneuvered *Atlantis* to a point in the ocean some thousand miles due east of Washington, D.C., and from this point he proposed to head north-northwest so as to cross the Gulf Stream at right angles. A hydrographic section was to be made across the stream and the ship would stop for temperature measurements and water samples about every six hours night and day.

This routine was begun just outside the estimated edge of the Gulf Stream and carried on without difficulties for three days. On August 21, however, at 3:00 A.M., the winch was started and it was noticed that one end of the drum was badly cracked and out of line. "This is our worst catastrophe so far," wrote Iselin, whose own interests lay more with this hydrographic work than with biology.

There is no way of continuing with the hydrographic work and making a decent job of it. . . . We could perhaps continue the section, going down to only 1,000 meters [instead of 3,000], but this would not be very valuable and we would run the chance of putting the winch completely out of business so that Clarke could not finish his light-intensity work. After thinking things over, we gave up the section and headed northwest for the colder waters just outside the edge of the continental shelf. Since we can no longer use the trawl boom for deep-sea towing, as it is badly bent, and since our hydrographic work is stopped, there remain only the light-intensity and plankton observations, and we will probably get to Boston a few days before schedule.

Few were disappointed, besides Iselin and Zorell, as *Atlantis* swung farther to the west and sailed for home. Within a single day the signs of approaching land multiplied — a tramp steamer, the end of the sargasso weed, the onset of cooler weather.

On August 22 a strong breeze came out of the northeast, and *Atlantis* boiled along at ten knots. She made such good time that it was considered prudent to heave to for the night rather than risk fetching up on Brown's Bank off Nova Scotia in the middle of the night.

"It is a great pity to have to waste such a fine fair wind. It would carry us to Boston in twenty-four hours . . . [but] we still have Clarke's plankton migration observations to make."

The wind held steady the following day and Iselin let *Atlantis* run into the Gulf of Maine. He hoped that the weather would moderate by evening so that Clarke could make, at a single station, a series of plankton hauls and light-intensity measurements that would extend around the clock. The dozen or so measurements he had already made showed that sunlight penetrated to depths of approximately 130 feet in northern waters and almost four times as far in the clear waters of the Sargasso Sea, where there was less plankton and other suspended matter.

Clarke hoped to relate these differences in light penetration to the vertical migrations of plankton, but this work needed a more controlled and restricted survey and was begun a year later in the Gulf of Maine and Woods Hole Harbor.

Monday, August 24. We hove to at 4:30 A.M. and by five had started our first series of observations. We made a haul every hour until nine. . . . Meanwhile it had clouded over and the wind had begun to freshen from the northeast. The barometer . . . started down. By ten o'clock, when we started to lower the photometer, it was blowing 25 m.p.h., and we could not keep the cable from straying out badly. At the noon plankton hauls we began to smash our gear and rip our nets. However, we finished a double series down to 130 meters. At that point I went to bed, having been up since five and having had only one hour's sleep last night.

Clarke, left in charge of the operation, continued the hourly hauls. The wind rose and *Atlantis* began to pitch and heave so wildly in the steep gray seas that handling the nets from the swaying platform became a dangerous business. At 6:00 P.M., with a cold rain coming down in sheets upon the deck, two nets in a row were smashed, and the twenty-four-hour plankton series came to a premature end.

"To think of all the moderate days we let pass hoping for a flat calm for this work, and here we have been trying to make the observations all day in a storm," wrote Iselin.

Atlantis hove to again that night and at eight the next morning set out for Boston, some 150 miles away. All day she rolled and pitched through a rough gray sea, and when a cloudy evening faded quickly into night, no one aboard had sighted land. In the bow of the ship, a seaman in oilskins and seaboots was on lookout duty. While he watched for the lights of coastal vessels and for those that should mark the approach to Boston, he could dimly hear the shouting and laughing of a party aft, where the last of the liquor that could not be brought in under Prohibition was being put away.

At midnight Thatcher's twin lights off Cape Ann were raised ("and so much for Old Physic Face's navigation"). The sailor at the wheel got the first faint and curiously unnerving smell of bayberry and pine, and one of the partygoers daringly hung over the side of the ship and painted out Clarke's porthole to trick him into sleeping late. *Atlantis* changed course and headed southwest toward the Graves and Roaring Bulls.

The ship never entirely quieted down that last night out. A few still awake heard a whistle buoy pass abeam in the hours before dawn, and three or four watched a cheerless gray light spread across the sky. And by that pale light they saw land. Off the starboard bow Deer Isle came out of the early-morning darkness, its penitentiary buildings barely discernible and its rocky shores, exposed at dead low tide, giving off a strong shore smell.

By seven most of the men were on deck (including Clarke), and twenty minutes later the ship came up into the wind off City Point and the starboard anchor was let go. Clowser ordered the ship run astern to see if it would hold, and found, to his dismay, *Atlantis* gently grounded on a mudbank. The maiden voyage was over.

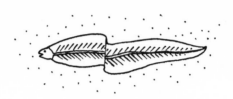

2 Hurricane in the Gulf Stream

Mister, the trouble with these small vessels is that they scare you to death long before they drown you.
— Comment made to Columbus Iselin by a man in Shelbourne, Nova Scotia

Atlantis was in Woods Hole. She was made fast to a wooden dock that extended from the Oceanographic Institution's new three-story brick building out into the sparkling waters of Great Harbor, and it was now up to Henry Bigelow, the Institution's director, and Columbus Iselin, the ship's master, to decide how best to use her.

The ship's schedule would be shaped by the objectives of the Institution, and Bigelow believed that over the next eight or ten years his staff should try to undertake enough biological, physical, chemical, and geological investigations to form a small but well-balanced body of fundamental oceanographic knowledge. He further believed, shrewd Yankee that he was, that it made no sense to plan such investigations until each scientist had gone out on *Atlantis* to see for himself the conditions under which he and his equipment would have to work. Only then could the eight or ten scientists and their students and assistants plan the yearly cruises.

"Yearly" is somewhat misleading. The Oceanographic had been established solely as a summertime institution and almost everyone on its staff worked at a college or university during the academic year. The vessel would typically be used each summer for short voyages and would then be free to make a long cruise to warmer waters in the winter if scientists could be found to use her.

The first short voyages undertaken by *Atlantis* began within a month of her arrival in Woods Hole and, as Bigelow had predicted, were partly sea trials to test the changes made in ballast and rigging and

47

In 1931 the Woods Hole waterfront was dominated by the old fisheries building, left, the laboratories and library of the Marine Biological Laboratory, center, and the brand-new Woods Hole Oceanographic Institution. All front on Great Harbor, which opens into both Martha's Vineyard Sound and Woods Hole. The Hole is the passage from the sound into Buzzard's Bay. *(John H. Welsh photo.)*

partly trials of the scientists and their gear. On cruise 4, for example, the stated objectives were to see if Alfred Redfield, a Harvard physiologist and senior member of the Oceanographic, got seasick (he didn't) and if he could catch some deep-sea fish.

Late one November afternoon, *Atlantis* boiled out of Woods Hole before a stiff offshore breeze and by the next morning had passed the edge of the continental shelf and was sailing over deep water. It was a bright, brisk fall day. When Redfield came up on deck, he noticed a considerable number of migrating land birds that had been blown out to sea by the offshore wind — juncos, starlings, sparrows. As he watched, some caught up to *Atlantis* and hopped down upon her deck to rest; others dropped into the sea behind.

Harold Backus and a loaf of stale bread appeared with the first of the

Henry Bigelow, left, first director of the Oceanographic Institution, and Columbus Iselin on *Atlantis* just after the ketch came into Woods Hole. *(John H. Welsh photo.)*

birds. The gruffly good-humored Yorkshireman, who had decided to move his family from England to Woods Hole when appointed chief engineer, had an abiding love for birds and had already begun to feed those that landed on the ship. He was building a few cages too, and, somewhat like the imaginary Mr. Glencannon of *Inchcliffe Castle*, who put a cat on blotting paper in his engineroom to see if the animal would sweat through its feet, Backus wanted to see if a sparrow and a junco might produce a sparco.

By ten in the morning a two-meter net had been rigged and swung over the side. It was allowed to trail behind the ketch on 1,600 meters of wire until early afternoon and was then brought back on board. The catch was disappointing — eight fish mangled beyond recognition and fragments of bright orange prawns. Deciding that the animals needed protection from abrasion, Redfield and Iselin rigged an empty lard pail in the cod end of the net and sent it down again. Since the skin of deep-sea fish is thin and delicate, however, the second haul was also badly abraded. Several prawn survived the trip in good condition, though, and Redfield managed to keep them alive for a short time.

After one more of these short trips, *Atlantis* set off on her first winter cruise. Iselin wanted to run a line of hydrographic stations through the southern portion of the North Atlantic to complement the more northerly section made during the maiden voyage. It was a long cruise — almost three months — and although one of the officers, convinced that he had appendicitis, quit as the ship crossed the lonely waters of the South Atlantic and furious rows erupted among the crew, the running of the hydrographic section became so efficient and routine that Iselin began referring to it as "the old grind."

The newly repainted ketch, now gray, behaved reliably with her additional ballast, and her scientific equipment worked smoothly. Her stove and plumbing were in order. With Terry Keogh and the other college friends gone, the crew had settled down. It was a successful cruise, professionally run. Only Clowser was disappointed.

First Officer Clowser, still wishing to be back with Cunard and still used as a scapegoat by the ship's company, had assumed command of the ketch halfway through the voyage. He had brought *Atlantis* north without incident, but at the end of the cruise, in guiding the ship up the Acushnet River to clear customs at New Bedford, Massachusetts, he passed to the wrong side of a buoy and ran aground. Although the ship floated free on the rising tide, Clowser, who had put the ship on a mudbank before and who had also run her smartly up onto the cap log of the Institution's dock, did not dare tell Bigelow that he had misread a channel marker. Instead he phoned the director at his office that cold spring day and told him that he had hit an uncharted rock.

Bigelow hit the ceiling.

"*You* hired him," he said to Iselin as soon as he had stomped down the hall and found the man. "Now you fire him!"

Clowser's departure could not have come at a more awkward moment. The Institution was about to begin its second summer season but its first fully functional one using *Atlantis,* and a reliable skipper for the new research vessel was a necessity.

Bigelow, who thought of the vessel as Iselin's responsibility, sent the young man hurrying down to South Street on New York City's waterfront, and there, with the help of a merchant master who knew most of the city's seafaring men, Iselin met Captain Frederick McMurray.

McMurray was a man in his fifties who was on the beach for grounding a freighter. Although born in upstate New York, far from the water, he had put to sea at the age of thirteen as mate, cook, deckhand, and general factotum on the sloop *Wanderer.* Upon deciding to make the sea his career, he had attended the New York State nautical school on the old square-rigged school ship *St. Mary's,* shipped out around Cape Horn on *St. David,* served on the armed transport *Harvard* during the Spanish–American war, and worked aboard the U.S. drydock *Dewey* as that mammoth structure was towed halfway round the world. Later he had sailed on mail steamers in the Orient, including one, the *Zafiro,* that carried the beautiful girls of the Bandman Opera Company between Hong Kong and Manila. Later still he had been sailing master of the nonmagnetic research vessel *Carnegie* and commanding officer of the school ship *Newport.*

McMurray's last job had been with the Isthmian Line. For nine years he had commanded a number of their freighters until, through an unfortunate error in navigation, he stranded *Steelmaker* on an atoll in the South Pacific for thirty-six days. In those days many steamship lines had a rule that if you touched bottom you were fired, and although McMurray was eventually reinstated to command, the line took so long to settle the matter that the captain left its employ in 1930. He was just about fifty years old then, a gruff man with more sailing experience than anyone had use for, and he was having one devil of a time finding a ship.

When Iselin met and subsequently hired McMurray in the spring of 1932, the captain was a balding, sharp-eyed man with the beginnings of a potbelly and a pipe that resided permanently in one corner of his

Captain Frederick McMurray conning *Atlantis* into port, always an anxious procedure. *(Jan Hahn photo.)*

mouth. Like most shipmasters, he was a lonely man, and he enjoyed puffing away on his pipe and telling without much emotion of the times he had rounded the Horn as an able seaman. His memories of furling *St. David*'s headsails as seas washed over her bow were vivid, and it had not been unusual, he said, to spend two or even three weeks straight in wet clothes before coming into port ripe for any frolic that might come his way.

In the spring of 1932 McMurray left New York City, his friends, and a wife — all without apparent regret — and moved to Woods Hole. He first took *Atlantis* to sea in June for the Bureau of Fisheries, whose biologists were studying mackerel eggs and larvae. By July the regular work of the Institution had begun, and the ketch was used to extend the light-intensity measurements and plankton studies that had previously been made on her maiden voyage and to supply bacteriologists and chemists with samples of mud and water.

In August the physical oceanographers had their turn, and a cruise was made to study the Gulf Stream. Iselin, being a physical oceanographer, was particularly interested in this strong current that, although recognized since the early 1500s as the great sea mark of the east coast of North America, had not been adequately described, much less explained. He hoped that by sailing *Atlantis* across the stream at varying angles — especially from Nova Scotia to Bermuda and from the offing of the Chesapeake Bay to Bermuda — and by sailing these routes at different times of year, he and his colleagues could gain some idea of the Gulf Stream's seasonal fluctuations in position, size, and speed.

So it was that on an unusually cool summer's day in August 1932 *Atlantis* let go her lines and moved out of Great Harbor. The Nonamesset bell buoy passed abeam, and upon orders from the captain, all sail was set and the vessel filled away up Vineyard Sound. All during that long afternoon the ship slid past the shoals off Martha's Vineyard and Nantucket — Cross Rip, Halfmoon, Handkerchief, and Stone Horse — and by evening was being whisked by the tide through Pollack Rip Channel. The ketch sailed northeast all the next day until, off Cape Sable, Nova Scotia, she turned toward Bermuda and began heaving to for stations.

The first hydrographic station was made at 1:00 A.M. As the ship's deck lights cast their wobbling reflections on glassy black swells that passed beneath the ship, the mate on watch took up his position on a small platform that extended from the ship's starboard side just forward of the wheelhouse. From here he could reach the hydrowire that ran off its own winch into the dark water beneath him, and to this wire he attached eight Nansen bottles, one by one. Two sailors helped him, one working the winch and the other bringing him the bright yellow bottles from their racks in the deck lab. When all the bottles had been lowered over the side of the quiet vessel, each hanging at a predetermined depth, a bronze weight called a messenger was slid down the wire. The weight caused each bottle to turn end for end in a motion that both closed it, thus securing a water sample, and triggered the reversing thermometer mounted on each bottle to register the exact temperature of the water. The bottles were than drawn up and one by one delivered to the deck lab, where one of the scientists read the ther-

With one foot on what became known as the "hero platform" and the other braced against the ship, Arnold Clark detaches a Nansen bottle from the hydrographic wire. *(Courtesy Woods Hole Oceanographic Institution.)*

mometers and drew off water for chemical tests. Some of the tests were carried out in the ship's labs, but most, including the all-important determination of the water's salinity, were done in Woods Hole.

The second station began when the watch changed at 4:00 A.M. and was finished just as the sky began to lighten. A third was made during breakfast, a fourth at 10:30 A.M., and still another at 3:25 P.M. The last for the day was begun shortly before midnight.

"Light air, smooth sea, fine and clear," wrote the first mate in the log before turning in for a few hours' sleep.

Twenty such hydrographic stations were made on the way to Bermuda, all under the direction of a chemist named Dick Seiwell. Seiwell was a quick, intelligent young man, but in Bigelow's estimation carried a chip on his shoulder that might make it hard for him to get along with the others on the ship. He was hired on a temporary basis, therefore, and sent to sea almost every time *Atlantis* made a section across the Gulf Stream. As expected, he proved to be an agressive man, always pushing himself and his colleagues hard and wringing the last decimal place from each temperature measurement he made, but he was accepted without complaint aboard the ketch.

On August 23 *Atlantis* stood into the harbor at St. George in Bermuda, and once the ship had cleared customs and received her pratique, Seiwell and most of the others left her in the care of Captain McMurray and rowed ashore. Crew and scientists stuck together in these early days, and having ranged along this quiet waterfront before in search of a good feed and a bottle of rum, the ship's company headed straight for the White Horse Inn. Within the hour several of the Scandinavians had begun to sing and First Officer Tom Kelley made a bet with himself that Seiwell would eventually do one of "his crazy things."

For the rest of that hot summer's day the men from *Atlantis* amused themselves in the tiny village of St. George and toward evening drifted back to the ship in threes and fours. There was a rule on *Atlantis* that everyone had to be back on board by eight each morning. The sailors turned to then and did their cleaning, painting, and repairing. Once their jobs were done, they were free to go ashore again. Since the crew was allowed to drink on board when the ship was in port, it had become one of Kelley's routine jobs to order a fifteen-gallon keg of Barbados rum to be divided among scientists and crew.

On this trip, Kelley cleared his purchase through customs as usual, rowed the keg out to the ship, and only then discovered that by mistake he had been given a keg of uncut, undrinkable, 195-proof alcohol. Since it could not be returned through customs, Kelley and Seiwell (who, it was said, "could make booze outta nothin'") borrowed a recipe for rum and purchased the necessary gallons of prune wine, brandy, and glycerin.

Seiwell liked to work at night, and he liked to be hot. That night he turned on the lights in the lower laboratory, turned on the gas burners to heat some of the mixture up, and stripped down to a pair of under-shorts. After clearing the counters of glassware, chemicals, and the tops of human skulls that he used as ashtrays, he and Kelley sweated through part of the night pouring gallons of distilled water and prune wine into the alcohol and then tasting, remixing, and tasting again. At last they hit upon a mixture as good as Barbados rum, and pouring it out of wax-lined specimen barrels into bottles and jars, found they had thirty-five gallons of rum, more than twice the usual amount. There was a wild celebration in the steaming, perfumed lab that night, and someone, getting hold of Seiwell's dog, Buttons, used black ink and Mercurochrome to draw a macabre approximation of the dog's skeleton on his brown-and-white coat.

Several days later, after Chief Engineer Backus had corked his remaining bottles of rum and dropped them gently into their accustomed hiding place in the ship's fuel tanks, Atlantis left Bermuda and headed back across the Gulf Stream. Again hydrographic stations were made night and day, at first with a single cast of eight water bottles and then with two casts of eight bottles each in deeper water. Care was taken not to hang the lowest Nansen bottle anywhere near the sea floor, where it might hit the bottom, as it had done at least once before. Since Atlantis had no sonic sounder, and since sounding with lead and line was too time-consuming in deep water, the widely spaced and none-too-accurate depths on navigation charts had to be used. The water bottles were hung on the assumption that these measurements were as much as a thousand fathoms off.

During the first few days of September, Atlantis spanked along to the west until she reached the mouth of the Chesapeake; then she changed course and headed home. For several days the ketch reached gracefully northward in a fair breeze, but the wind began to increase and haul ahead and by September 7 it was blowing hard from the north. McMurray began to mutter, and true to his gloomy predictions the barometer started down. For no apparent reason a small collection of houseflies, land birds, and butterflies gathered on the ship.

By early afternoon on September 8 the seas had made up into great mounds of glittering gray water, but it was still just possible to make a hydrographic station. The sailor carrying the water bottles slid wildly back and forth across the wet deck from rail to deck lab, and the mate fastening bottles onto the wire stood in swirling water up to his knees on every other roll.

The station completed, *Atlantis* started on again, laboring hard, and almost at once McMurray ordered her hove to and the mizzen furled. Running under a single headsail and storm trysail, the ship went off on a starboard tack and through the long gray afternoon and evening she pitched and rolled as the wind rose and the seas built.

"September 9, 2:30 A.M. Mod[erate] gale, rough sea, overcast," read the log, yet at five in the morning Seiwell and the crew tried to make another station. McMurray came up on deck and ordered the jumbo furled. The sailors almost had it down before the wind gave it a last violent shake and split it open along a seam.

"Wind and sea increasing to a howling gale," wrote an unsteady hand, and at 6:00 A.M., "Hove to on a port tack to make an offing. Vessel laboring heavily and straining." An extra man was put on the wheel.

The wind was soon exceeding gale force, and seas heavily streaked with foam rose into the gray sky and toppled and crashed around the ship. Above, a layer of low, ragged clouds, torn from the east, went flying across the sky, and below, the churning crests raced across the sea. Wind shrieked through the rigging, rain drove across the decks, the ship creaked and groaned, waves roared by like trains, and everything within sight, sound, and feeling seemed bent on hurling itself over the edge of the world.

Still the wind rose and the barometer dropped. McMurray, braced in one corner of the wheelhouse, held a cold pipe upside down in his teeth and at 10:00 A.M. ordered the engine slow ahead to keep the ship's head into the solid wall of foam and spray that marked the wind's direction.

"Wind blowing with hurricane force," someone managed to scrawl in the log, but just how fast it blew could not be logged, for the needle on the hand-held anemometer had blown off the scale.

At noon McMurray ordered the storm oil tanks turned on, and now trickles of vegetable oil came leaking out around the ship's bow. Kelley, who wanted to see if the slick would really work to windward and calm the surface turbulence of the seas, as so many seamen claimed it would, climbed up on deck through the forward companionway but could see nothing through the spray that shot across the ship. From handhold to handhold, Kelley ventured out upon the deck, which seemed to drive upward, then fall out from under him with premedi-

Atlantis, under jumbo and main trysail, takes one of her spectacular rolls. *(Courtesy Woods Hole Oceanographic Institution.)*

tated violence. He felt his way aft as far as the mainmast, then a step at a time climbed the rungs on the mast into the full force of the wind. For six or eight steps Kelley could neither see anything through the blast of spray that stung his hands and face nor hear anything above the roar of the storm, but when he had climbed a few feet higher on the swaying mast he found he had left the spray below and could open his eyes. He clung there for a moment, arms around the mast, and looked out across the most savage battle he had ever seen between wind and sea. To windward he could barely discern a slick that in some small measure was smoothing the superficial wrinkles on the mountainous waves. So much for vegetable oil.

When Kelley went below, his battered oilskins running like the rain itself, he found to his surprise that McMurray was in better spirits. The storm was at its worst. *Atlantis* was having as hard a time of it as she had ever had, but the barometer had ceased to sink.

"Can't get worse," McMurray said, and by eight that evening the sky was beginning to clear in the west and at midnight the sturdy *Atlantis* began to make some headway against the seas.

"*Atlantis* went through the whole of that storm without damage of any sort," wrote Bigelow several days after the ship had returned to Woods Hole. "It is a comfort to know that they can go through a first-class gale without difficulty."

Using the data gathered on this cruise and on two or three earlier ones, Iselin wrote the first of his many papers on the Gulf Stream. At this point he was only trying to describe the Gulf Stream, or, as he put it, "delineate its structure," but he could already see that it had three sections, each with different characteristics. He chose to call the southernmost and relatively shallow portion the Florida Current, the deep, compact stream that flowed from the east coast of Florida to Nova Scotia the Gulf Stream, and the ill-defined, branching portion that wandered east toward Europe (and about which little was known) the Atlantic Current.

While Iselin worked on the circulation of the western North Atlantic, Seiwell concentrated on analyzing the chemical properties of the Gulf Stream and adjacent waters. He was particularly interested in measuring the seasonal cycles of such important nutrients as phosphorus.

Two months after *Atlantis* returned from her rousing trip to Bermuda it was decided to sail the ship around to the Electric Boat Company on the outskirts of New London, Connecticut, to have a Fathometer installed. The Submarine Signal Company made these sonic sounding devices, and the one installed on *Atlantis* had two modes of operation. In shallow water, to a depth of 125 fathoms, an oscillator produced a sound somewhere between a *cheep* and a musical tone, and as these sounds were bounced off the sea floor, the depth was registered automatically. In deep water a different oscillator produced a more powerful sound, like that of a sledgehammer being banged against the hull, and a person acting as receiver had to listen for the echoes through earphones and compute the depth with the aid of dials and flashing lights. This procedure was hardly automatic and was further complicated by the necessity of stopping the ship's engine in order to hear the faint echoes. Still, it was a great improvement over a sounding lead and would enable physical oceanographers to distribute their water bottles more effectively and geologists to map many of the unfamiliar structures of the sea floor.

Iselin had gone along on this short cruise to New London to oversee the installation of the Fathometer, and when the ship left the Electric Boat Company he facetiously remarked that he detected a slight list to port. He had noticed this list before, he said, and it seemed to get more pronounced each time *Atlantis* visited a shipyard. After discreet investigation Iselin concluded that a process he called the "Backus factor" was at work.

The Chief, as the engineer was usually called, had taken it upon himself to see that the Institution was not robbed by shipyards, whose rates he considered exorbitant. He corrected this injustice by gradually collecting a complete set of tools.

"You know, Skipper," he told Iselin in his Yorkshire accent, "when the whistle blows, these shipyard fellows drop everything and run for the gate. I simply pick up the best of what they leave behind."

It did not take cost accountants at the Electric Boat Company long to figure out how the Backus factor worked, and they began adding a surcharge to the bill for each day that *Atlantis* was in the yard. "When I pointed this out to Harold," continued Iselin, "he was outraged, [but] in the end, there was not a tool that he did not have in all sizes and designs."

With the Fathometer installed and the Chief's tool collection supplemented, *Atlantis* returned to Woods Hole. She was scheduled to make only one more cruise in 1932, to Bermuda. She ran out and back to the islands, bucking her way through a succession of storms, but most of the way managed to make deep acoustic soundings once each hour and shallow soundings almost continuously. As she was returning to Woods Hole along the New Jersey coast in bitterly cold weather, hoping to make port by Christmas, the new Fathometer suddenly indicated a rise so abrupt and unexpected that the ship was turned around and the depth checked with the lead. The two methods agreed, and it was discovered that the ship was riding above an old wreck.

Atlantis slid past the Elizabeth Islands and into the icy harbor at Woods Hole on December 23, having completed fourteen cruises in her first full year at the Institution. As was the custom, a Christmas tree was hoisted to the top of the mainmast, and in the cold, gray Cape Cod evenings the half-dozen people who might walk along Water Street past dark buildings closed for the winter could see the tree with its colored lights gently rocking above the ship.

3 Caribbean Cruises

Being in a ship is being in a jail, with the chance of being drowned.
— Samuel Johnson

Toward the end of January 1933, a great coming and going broke the usual quiet in the village of Woods Hole. Trucks and cars came rattling down Water Street to deliver food and equipment to the Institution pier, and in the main brick building several scientists and technicians hurried to collect the glassware and reagents they would need on *Atlantis*. The first of a series of long winter cruises to the Caribbean was getting under way, and this one, like the three that followed, was being jointly sponsored by the Woods Hole Oceanographic Institution and the Bingham Oceanographic Foundation at Yale University. The foundation had agreed to contribute funds, scientists, and equipment in return for working time aboard *Atlantis*.

The general plan was to make a hydrographic survey of the Caribbean and Gulf of Mexico, exploring a new part of this territory each winter. Studies of fish, plankton, and sediments would be added as time allowed.

On the last day of January, Columbus Iselin and Alfred Redfield came down from Harvard to join the ship, and on the first of February, *Atlantis* steamed into Vineyard Sound.

"Some of the scientists sick," wrote First Officer Kelley in his log as the ship headed south.

By the evening of February 3 the ship was sailing across the mouth of Chesapeake Bay, and all around her the lights of fishing vessels rose and fell on the long, dark swells. A lookout had been posted in the bow, as usual, and in the early hours of the morning he shouted to the mate that he thought he could see a column of smoke dimly silhouetted against the sky. The helmsman was directed to change course and "at four A.M. we came to *Victor of Nantucket* burning at sea."

Around the fishing vessel low flames cast a glassy circle of re-

Albert Parr aboard *Atlantis* during one of the cooperative cruises. Fish and sargassum are being kept in temporary aquaria on deck. *(Courtesy Woods Hole Oceanographic Institution.)*

flected light, and at its cold perimeter bobbed an open dory lit by a lantern. *Atlantis* cautiously approached the open boat, and once she had picked up the five-man crew, she reversed her course and delivered them to a fishing vessel that would take them ashore.

The ketch resumed her run to the southeast. In a driving snowstorm she crossed the Gulf Stream, heaving to every couple of hours to make a hydrographic station, then headed for Bermuda, where several scientists from the Bingham Foundation were to join the expedition.

Chief among the new arrivals was Albert Parr, a handsome Norwegian scientist and curator of the Bingham Foundation. He would be chief scientist on the cruise after Iselin left the ship in Nassau.

Parr had brought onto the ship with him a gigantic trawl of bizarre design with which he hoped to catch deep-sea fish. He was especially eager to sample the larger, faster animals that, if they existed at all, had so far avoided capture in small, conventional nets. Parr's trawl had a triangular opening 50 feet on a side and was 128 feet long, almost as long as the ship herself. With its three weighted trawl boards, the net weighed more than a thousand pounds.

On February 18, as *Atlantis* sailed steadily toward Nassau, the trawl was tried for the first time.

"Rigged big net during afternoon," wrote Kelley. "Caught very little, but net worked OK. Quite a job getting it aboard."

On the twentieth it was tried again — lowered into a calm blue sea on 1,000 meters of wire at four in the afternoon and left there until five the next morning. The winds remained light and variable all through the night, and waves hardly moved the ship as the big winch ground and grumbled, reeling in the cable. In the gray half-light before sunrise the trawl boards that held the mouth of the net open broke the surface, and without too much difficulty they were hoisted up and swung dripping onto the deck. Sailors then began pulling the net in hand over hand. They heaved and hauled on the wet cords and soon it became apparent that the cod end of the net was full of fish. Peering over the rail, the exuberant Parr tried to remain calm. He urged the men to take care, to go slowly. Cautiously, then, the net was lifted over the rail and its fragile contents landed on the deck.

Parr was in ecstasy. Out of the net and into enameled sorting trays came 491 fish of 47 varieties. There were black fish with rows of lights down their sides, angler fish with fleshy appendages growing from their heads, fish with fierce fangs, light fish, dark fish, small fish, relatively large fish, and a squid with eyes on stalks that hung out of his head like a couple of grapes. Twelve of the species had never been captured before and one fish was so entirely different from its next of kin that a whole new subfamily was set up for it.

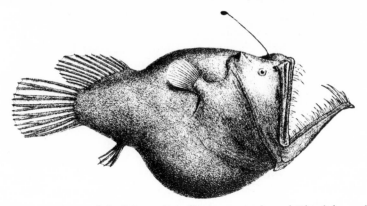

Melanocetus murrayi, one of the fish caught in Parr's oversized trawl. This fish was first discovered on the *Challenger* expedition, 1873–1876, and named after one of her scientists, John Murray. *(From* Challenger Report *22, pt. 57.)*

"Net aboard. Beautiful catch. Headed for Nassau," wrote Kelley.

Using power more often than sail to make way against southerly winds and towing a small surface net called "the spinach collector" to catch sargasso weed, *Atlantis* sailed on to the south. The big trawl was tried once more, but either the weight of the catch or some great thrashing creature ripped the net completely off its frame, and as the ketch carried no spare, the project was reluctantly abandoned.

When *Atlantis* arrived in Nassau, Iselin and Redfield left the ship as planned and Parr took charge of the scientific program. Under his direction the ketch sailed slowly through eastern portions of the Caribbean and in the course of the next three months made well over one hundred hydrographic stations. In May she began working her way north, and although she had logged nearly 10,000 miles on this cruise alone, Parr's projects had scarcely begun.

On the night of May 22 *Atlantis* again slid past the bell buoy off Nonamesset, which, with its hollow and asynchronous clangs, marked the entrance to the Hole. McMurray had ordered the sails lowered, the engine started, and the lights trained ahead as *Atlantis* moved carefully in through patches of chop, glossy slicks, and whirlpools that, except at the very moment of slack tide, express the intricate constrictions and configurations of the rocky harbor floor. Off Ram Island, discernible only as a dark shadow off the port bow, the ship slowed still further, swung to starboard in a gentle arc, and finally, avoiding the lobster pots whose striped buoys winked and bobbed in the ship's lights, eased into her berth before the Oceanographic. The ship's cat, Mitzi, made a spectacular leap for the dock and, as was her custom, waited for the ship to tie up before springing back on board.

"4:00 A.M. All fast to dock."

After breakfast a small crowd of friends came waving and calling onto the dock, while up the street, across from the Woods Hole Inn, a self-employed local lady who was held in high esteem by seamen, and who had called Iselin to find out when the ship was due, waited to pick up her sailor friends in her shiny black four-door car.

But there was no vacation for the crew after their four months at sea, and within a day or two they began to clean and repair the ship in anticipation of the summer season. A new suit of sails was bent on, whaleboats replaced the dories as lifeboats, and a radio transmitter and receiver, lent by the Coast Guard, were finally installed.

A year later, in 1934, *Atlantis* made a second and much shorter cooperative cruise to the Caribbean. This time the collection of sargasso weed and flying fish was added to the hydrographic program, for Parr had become interested in determining the source of the several varieties of weed he had collected on the first cruise. In addition, one of his assistants wanted to collect flying fish (creatures that McMurray insisted resembled grasshoppers). During this cruise droves of the little fish were seen skittering away from *Atlantis* as she sailed south over the barren and intensely blue waters of the Sargasso Sea. At night the fish were attracted to the ship by her lights, and they would inadvertently fly against the hull or up on deck. One hit McMurray right on the forehead.

On this two-month cruise Parr and his colleagues explored the western portions of the Caribbean. They sailed to the Panama Canal Zone, then worked back over a vast coral plateau east of Honduras where the water looks alarmingly but deceptively shallow and where the sport fishing is superb. After making many hydrographic stations among the groups of small islands and banks there, they headed for Woods Hole by way of Miami. A Florida paper reported that the scientists had collected as pets three-pound marine frogs, a boa constrictor, iguanas, and other "repulsive-looking reptiles."

The *Washington Post* quoted one of the scientists as saying in part:

> We're home from the seas,
> A most studious crew

Of data we bring back a mass-o.
We sailed at our ease,
Having little to do
But study what's in the Sargasso.

Each morning we cruised
With a diligence great
On a ketch we were never afraid of;
We were never confused
As we looked deep and straight
To observe what the strange sea was made of.

We peered and we peered
With an all-seeing eye
As we sailed through the waters so tropic,
And the data we booked
Was important, oh my!
And most of it quite microscopic.

The third cooperative voyage to the Caribbean began in January of 1935, and although the scientific program was about the same as on previous trips, troubles began to develop with both the equipment and the people aboard *Atlantis*.

"You are the final authority for all that goes on whether at sea or in port," Iselin had written in McMurray's sailing orders. "You are also in charge of the scientific party. He [Parr] will understand that in matters of discipline and behavior, the scientists must conform to your wishes."

But they didn't, and by mid-February, after a month at sea, McMurray realized that for all his authority, he could not keep one of Parr's assistants from tangling with Spencer Greenwood, the radio operator, nor could he assuage Parr's own anxieties over symptoms that suggested that the chief scientist was beginning to suffer a relapse of a chronic and uncomfortable stomach disorder. And what could McMurray do to help Eric Warbasse, a young man with incurable cancer, when his forced humor got on people's nerves, or when he had trouble sleeping and Greenwood, already tired and irritable, stood Eric's night watches as well as his own?

There was tension aboard *Atlantis*, and as the vessel sailed toward the western tip of Cuba, McMurray hoped that the remaining hydrographic stations could be made before trouble erupted. The ship was out of fresh food and low on water, and Kelley had already drawn off and divided five gallons of the rum purchased in Bermuda, a clear indication that the crew was ready for shore leave.

Parr and McMurray finally agreed to make an extra port of call in

Tampa, where Parr and Warbasse could get medical treatment and where the steaming crew could blow their excess energy on festivities being held in honor of Gasparilla Day. Consequently, after the ship had made several runs north and south along what was known as the Tortugas line, *Atlantis* steamed up the low, flat coast of western Florida and at 3:30 A.M. on February 23 hove to off the bar. A pilot came aboard to guide them up Tampa Bay, and by 9:30 the ship was taking on oil and water at the Sun Oil dock.

Hardly had *Atlantis* moved to a more permanent berth when Terry Keogh rushed joyously aboard. With little persuasion he rounded the crew together in a waterfront bar and soon was telling them of the job he had just had driving a float in a parade for the Golden Jubilee of Cigar Makers. He had been drinking as he drove, he admitted casually, and when a heckler on the curb disparaged his float, he had shouted back, forgotten about the parade, and driven straight into the float ahead of him.

Terry went on to describe his attempt to sail a twenty-one-foot cutter around the world single-handed. From Woods Hole he had gotten as far as Jamaica and there had borrowed money on the boat, imbibed the proceeds, lost the cutter, and drifted on to a series of jobs in South America. He told of his exploits there — brave declarations of Irish ingenuity — and kept the men from *Atlantis* laughing and pounding on the bar late into the night. Once he had been so completely out of liquor and out of luck, he said, that he had stowed away on a freighter. Discovered, he had worked his way up from the lowest, dirtiest jobs to the position of third mate. But then the freighter had put in to port. He had gotten drunk while in charge of unloading the ship's cargo and had given an order that resulted in a railroad locomotive's being dropped over the side.

"Terry is here, very much so," wrote McMurray to Iselin after three days in port. "McLunin, Backus, and others of the crew got him straightened up. The last time I saw him he was quite sober and had gotten his clothes in good shape. As he does not take me very seriously, I am not in his confidence except when he thinks he is getting a rise out of me, so I cannot be sure of his movements. . . . It is reported that he left for New York on the *Millinocket* of the Bull Steamship Line last night."

On the afternoon of March 1, *Atlantis* moved slowly out of Tampa Bay, and few, if any, of the men on board saw Terry again. He continued to drift in and out of trouble, in and out of people's sympathies, until shortly before World War II. Still young, although not exceptionally strong, he had a job on a coaster then, carrying goods from Florida to New York, and one night, to escape the oppressive heat below, he slept on deck and was bitten by malarial mosquitoes. By the time his ship was off New Jersey, Terry was extremely ill. He was taken off by a Coast Guard cutter and transferred to a naval hospital, where he died.

After leaving Tampa, *Atlantis* sailed west through the Gulf of Mexico, making particularly rich hauls of animal life (mostly shrimp) off the Mississippi Delta whenever the half-rotted trawl nets that had been borrowed for the cruise held together long enough to land the catch on deck. After about a week of this work, when the ketch was shuffling along through an almost windless and oppressively humid night, one of the engineers discovered something seriously wrong with the engine. It was nearly midnight when he climbed up to the chartroom to get permission from McMurray to start disassembling the engine. As soon as it was shut down and the sails alone drew the ship along at a crawl, noisy laughter from a poker game could be heard coming up from the messroom. The poker players kept playing until the watch changed at 4:00 A.M. and the engineers kept working in the stifling engineroom until noon the next day. They put the machinery back together again, but it was not repaired. McMurray ordered all sail set, and without stopping for stations, *Atlantis* flapped restlessly north to the nearest port, Pensacola.

While the engine was being repaired in this muggy, low-lying port with its constant traffic of military planes overhead, the men from *Atlantis* occupied themselves in the usual fashion. By day they worked on the ship and at night they went to a movie or drank in a bar. One evening many of them caught rides to a restaurant and dance hall at Paradise Beach, some seven miles away, to attend a celebration held by the Coast and Geodetic Survey in honor of a survey just completed.

As usual, McMurray preferred to stay aboard, and when he retired at midnight few of the men had returned. During the night he was vaguely aware of the rattling gangplank and of footfalls on the deck overhead as the crew came back aboard in noisy twos and threes.

At 5:00 A.M. there was a sudden pounding on the captain's door and McMurray was abruptly awakened. There had been an automobile accident. William Potter, second mate, had been killed and two others from the ship were injured. Once dressed, McMurray hurried down the companionway, banged on Kelley's cabin, and told him to go talk to the coroner, while he himself drove directly to the hospital. There he saw Spencer Greenwood, the radio operator, who was in critical condition, and one of Parr's assistants, who was injured less seriously. With a Coast and Geodetic Survey man and two others, the three men from *Atlantis* had been driving back from Paradise Beach when an ensign from the naval station had come careening down the wrong side of the road and struck them head on. Greewood died that night.

Four days later *Atlantis* put to sea. Her engine immediately broke down again, and she was towed back to Pensacola through a heavy fog. There her engine shaft was replaced. On Sunday, March 24, the ship again steamed out of the bay. Just before supper the crew was mustered on deck and a short funeral service was held for Potter.

"At 1705 held funeral services over the ashes of William Delano Potter, 2nd officer of this vessel, who departed this life 16 March 1935," wrote an officer in the log. "His ashes were committed to the deep. 1712 — full ahead."

Atlantis continued her work, and on April 26 was back in Woods Hole. In spite of the restlessness on board, the engine troubles, and the sadness over the deaths of Potter and Greenwood, the scientific program had been carried out successfully. In fact, Parr had so many temperature measurements and water samples and so many jars full of sargasso weed, crabs, fish, mollusks, and shrimp, all of which needed careful examination, that he persuaded Iselin to let two years pass instead of one before launching the last and the longest of the cooperative cruises.

Cruise 65 began on December 27, 1936, and from the very beginning it was a bad-luck voyage. On a mild afternoon just two days after Christmas, the decorated pine tree atop the ship's mast was hauled down, the last of the stores were hastily brought over from the A&P food store, and the ship was towed out of Great Harbor against the tide.

Several very different kinds of work had been planned for cruise 65. Hydrographic stations were to be made during the first long stretch at sea; a cooperative program with the navy was to be initiated in Guantanamo Bay, Cuba; a geologist was going to try to obtain cores from a spectacularly deep trench north of Jamaica; Parr and his assistants were going to carry on physical and biological work once they boarded the ship in Kingston, Jamaica; and the Bureau of Fisheries had an interest in finding new shrimping grounds along the Gulf Coast. To fit all these projects into a single cruise, scientists and crew would have to work harder and faster than usual.

During the first week, *Atlantis* sailed south and southwest as fast as she could without stopping for stations. The weather grew warmer, the winds steadier. When the expedition arrived in the tropics, the work began and the fresh food ran out. There was nothing unusual in this. Fresh meat and vegetables were always rather abruptly replaced by canned foods, and evaporated milk was used in coffee, on cereal, and for drinking. But it *was* unusual to have the bread replaced with nothing but biscuits and cornbread, and after three days the crew began to complain.

The cook on cruise 65 was a stocky, well-built man in his early thirties. He had taken his turn at kitchen patrol in the peacetime army, and may even have done some short-order cooking before shipping out on *Atlantis,* but he was by no means an experienced cook. Part of his job was to order food for the voyage, and the Institution had made it clear that if he exceeded the eighty cents per day allowed for each man, he would be fired. Unlike many cooks that followed him, he was not bothered by this frugal policy: he took it to mean that he was simply to buy little, prepare little, and serve little. He stocked up on plenty of macaroni and canned fruit and took along only a small amount of yeast on the unlucky chance that he would have to bake.

The first leg of the journey — a leg that would last well over a month — had hardly begun when the bread ran out and the yeast went sour. Breakfasts then consisted of eggs, cornflakes, and corn muffins, and dinners were likely to be bully beef, canned vegetables, and cornbread. In the evening pieces of dry cake and more cornbread were put out for the night watch, and the cook was bluntly told that the roaches got to it before the sailors could and were welcome to it.

"We knew what we'd get for dinner before we stepped in the messroom," said one of the crew. "Garbage. Some of us would push the food around with our knives [most of the crew cut their meat with sheath knives], then shove the plate back at the cook with our compliments."

"So cook the goddamn stuff yourself," the cook would snarl as he slapped the uneaten food into a garbage pail.

When scientists, too, objected to the food, McMurray let the cook take hard bread from the ship's emergency supplies, but it was weevily.

In spite of the grumbling, which soon erupted into clattering rows between cook and crew, the hydrographic work went on and *Atlantis* sailed briskly south before the northeast trades. Within a few miles of Venezuela the ship turned northeast and worked her way along

In the warm climate of the Caribbean, life on *Atlantis* was growing increasingly informal. No more uniforms; no more remaining before the mast. *(Courtesy Woods Hole Oceanographic Institution.)*

the Windward and Leeward islands, many of which, with their exotic greenery and sprawling villages, were within sight yet tantalizingly out of reach. The mountainous slopes of Grenada slid by, then St. Lucia, Martinique (they could see the city of Fort-de-France), and Dominica. The days were perfect. The trades blew through clear blue skies each morning, and as the day wore on they piled clouds of fabulous shapes and colors up over the islands. It was fine sailing weather too, but *Atlantis* was always heaving to for another station just as she got fully under way. It was stop and go and stop and go, and would be for another two weeks.

"Because the cruise is so long, you are cautioned to keep a sharp watch on the crew's behavior and try to forestall any petty quarrels or acts of partisanship on the part of your officers," Iselin had written McMurray. "There has been a noticeable tendency for the ship's company to divide into two antagonistic groups. . . . Such disciplinary action as you may feel called upon to take will be fully supported by the Institution."

On January 30, just as McMurray thought he would have to put into effect the unspecified disciplinary actions that Iselin had suggested, *Atlantis* arrived at the U.S. naval base in Guantanamo Bay, on the southwest coast of Cuba, and the ship's company fell over themselves in their scramble to get off the ship. Bigelow had often told Iselin, and Iselin McMurray, that it was wise to let the sailors go on a binge for the first twenty-four hours in any port and only then to try to get work done aboard. Consequently, nothing moved on *Atlantis* until the first of February. Then, while the ship's biologists remained ashore sampling tide pools and seining portions of the town drainage ditch, *Atlantis* ran in and out of Guantanamo Bay, cruising alongside a destroyer that was attempting to make sonar runs on a tame submarine.

In the middle of February, with this part of the cruise completed and a geologist named Charles Piggot and his experimental corer aboard, the ship left Guantanamo Bay for Kingston, Jamaica. This short run would put the vessel in position to study the Cayman Trench, lying north and west of Jamaica. It was in the Bartlett Deep, the deepest portion of the trench, that Piggot wanted to test his corer. Columbus Iselin had assured him that *Atlantis* would be at his disposal for the better part of a week and that she carried sufficient wire to send his corer to the very bottom of the four-mile deep. With these aspirations Piggot and the others sailed down to Kingston in a single day, the crew groused about the food as if they'd never been off the ship, and another wild celebration took place when they cut loose in the uncomfortably hot but noisy and exotic port.

Unfortunately, the belligerence that had grown among the crew, exacerbated by cornbread, foul-tasting water, and long, boring days at sea, was too well entrenched to be dissipated in a single howling, dancing night. The next evening, as the cook lay in his bunk, he was assaulted by two of the crew.

"He narrowly missed being choked to death," McMurray wrote, "and from that time he has been a burden on me."

The last remnants of discipline in the galley fell apart, and McMurray quickly sent the badly shaken cook ashore for a week's rest. For the short trip to the Bartlett Deep he "replaced him with a regular bum" who could cook only when sober and who had brought along an inexhaustible supply of rum.

With this questionable addition to the ship's company *Atlantis* sailed out of Kingston on the morning of February 20 and by noon the next day was riding over 2,680 fathoms of blue water. The gun corer was sent down and, after a long round trip, returned with a mud core less than four feet long. At the next station the Fathometer indicated the much greater depth of 3,400 fathoms, almost four miles, and the ship's entire supply of wire was let out behind the corer.

"We missed the bottom."

To Piggot's intense disappointment, there was not enough wire on the ship. On a subsequent station the spooling device on the winch drum broke, and with no more work possible, *Atlantis* returned to Kingston for repairs.

Parr and his assistants now joined Piggot on the ship, and Parr, inadequately prepared for the rough living conditions by Piggot's mild complaints, was shocked by the cooking, the lack of hygiene, the smells, and the whole tense, sullen atmosphere.

After only two or three days out, the cook, who had also rejoined the ship, served spoiled meat, its age only partially disguised by onions and gravy. The meal was served on greasy dishes, and the forks that came to the gimbeled table had dried food still stuck between their tines.

"Where so many of the ship's company expose themselves to infection at every opportunity," wrote Piggot testily, thinking of the cook's hands covered with impetigo and the messman's oozing leg, burned on the galley stove, "it is imperative that all table gear be subjected to boiling water." But it was not.

Piggot also complained that the air that finally made its way into the stifling saloon came "in through the galley and over the swill bucket, and in the tropics it attained an unbelievable richness."

He was upset and annoyed, but Parr, a more volatile man by far, reacted to the stench as to a personal insult, and he was furious.

Living conditions had reached the flophouse level, he wrote, dashing off a letter as the ship approached Havana. "Administrative blindness" was responsible for the intolerable conditions. A few days out and essential provisions run short and the food becomes repugnant. Lumpy pillows and mattresses smell of their former occupants' distress

and the turkish bath atmosphere of the sticky cabins is periodically flooded with the vapors and gases of a privy which is in a condition to stop any normal metabolism through sheer discouragement and disgust. . . .

Despite vague assurances to the contrary . . . there has been a continuous decline throughout my experience with the ship. . . . When I contrast the happy good-fellowship of the first cruise in 1933 with the quiet, almost sullen association of many of the same men now, I can see very clearly what a great loss the ship has suffered. . . .

I am now considering the advisability of resigning my association in explicit protest against the conditions on board the *Atlantis* . . . .

While Parr mailed his letter and Piggot left the ship in Havana, McMurray took the cook and the messman to a hospital, where their infections were treated. "But as the men [later] went ashore and drank beer besides sweating in the hot galley, they nullified to some extent the value of the treatment."

One of the mates also appeared at the marine hospital, by orders of the Immigration Authority, and McMurray dourly notified Iselin that if the test for syphilis were positive, he would be deported at the Oceanographic Institution's expense. Two other cases of venereal disease were being treated on the ship (with argyrol and potassium permanganate).

"Kingston is responsible."

Sometime after March 8, when *Atlantis* had again put out to sea and was headed for Mobile, Iselin received Parr's letter.

"I'm shaking, I'm so mad," he told George Clarke, and apparently he told Parr the same by return mail. Iselin also wired McMurray, telling him to phone him immediately upon his arrival in Mobile.

In the meantime, Parr was composing another letter, this one severely and justifiably critical of a dilapidated current meter that had been put on board in Woods Hole with a half-dozen screws missing, a propeller blade badly bent, and all its gears frozen. It had failed completely on its first trial, had been fixed, and had failed again.

Parr was also struggling with miserable discarded trawls, lent by the Bureau of Fisheries, that "were so rotten we could pull them apart between our hands and not one of them came up again even once . . . being unable to lift the weight of their own contents."

While Parr was mailing this letter in Mobile, McMurray received the hot-tempered resignations of the entire galley staff and two of the original Scandinavian crewmen (who, having been in the United States for five years, had become citizens and were now free to find higher-paying jobs), and called Iselin to tell him that "there is nothing exceptional in regard to conditions aboard the ship. . . . Scientists are easily upset over ship matters, but they get over it."

Although Parr liked to dramatize his complaints, he was not just imagining a decline in conditions aboard *Atlantis*. The cost of operating and maintaining the ship had steadily risen as the scientific programs expanded, but the Institution's income had not. A fixed endowment yielded an inflexible $35,000 each year for the ship. Increasingly, economies had to be made. Some, such as paying no overtime to the crew, were inescapable. Others — the food — probably hurt more than they helped. But wise or foolish, the economies almost always affected the living conditions before they curtailed the scientific capabilities of the ship. Scientists didn't mind roughing it for a few weeks at a time, and even McMurray, trained on sailing vessels that had been driven too hard with too few men in unsuccessful attempts to compete with steam, was used to a wet cabin and cheap food. The crew, however, was less tolerant. When they learned in Mobile, for example, that unionized seamen were striking and that a strikebreaker could get $72.50 a month plus 60 cents an hour overtime, at least five left *Atlantis*.

The final cause of the decline was the laissez faire policy of Woods Hole, or "administrative blindness," as Parr called it. Both Bigelow and Iselin had every intention of running *Atlantis* safely and well, but they were too busy to see to the details and simply left them to McMurray. Bigelow, for example, was alarmed at the thought of weevils in the ship's emergency stores, so he told Iselin to tell McMurray to check regularly on the provisions as well as the firefighting equipment and so forth. Just as Bigelow realized that Iselin would not have time to do the checking himself, Iselin realized that McMurray wouldn't either. But a ritual compliance was expected, so Iselin wrote to the captain, half apologetically, "I know perfectly well that you have your own way of attending to such matters." And by and large, that was that for the next fifteen years.

"The sad fact remains that nobody in a position of authority at this institution likes administrative work," Iselin wrote Parr at the height of the cruise 65 flap," and until we find someone who does, we are never going to be particularly efficient in the running of our ship."

In Mobile, McMurray had the luck to find a Filipino cook and two messmen. The ship left for a week of trawling, and although on this leg of the journey Parr continued to complain of lost equipment and rotten trawls, he also noted that "a Filipino galley crew has caused the most amazing improvement in food, service, and cleanliness."

The new cook, who had quickly recognized the ship's refrigeration problems, was serving curries so hot that the scientists claimed, with tears in their eyes, that the food had to be washed down with formaldehyde. Unfortunately for McMurray, his constitution was unable to handle hot food, and he was soon blaming the Filipino for overstimulating his "bloody awful gasworks." The captain was also mildly disturbed by the cook's eccentricities; not that he minded his talking to the canary that he kept in the galley, but he didn't like it much when the man pointed knives at his back when he thought the captain wasn't looking.

As *Atlantis* left the Gulf of Mexico and headed for Miami, Parr wrote Iselin one more letter and told him, in a much calmer and more friendly way, that another reason for his intense regret over wasting time on an ill-equipped ship was that during his absence his daughter had been critically ill.

Iselin, himself a father by this time, replied that he too regretted the crises and inconveniences of "ill-fated cruise 65," and added that "when children can be so frightfully sick, I sometimes think we are all wasting our time fooling around the ocean."

Parr and his colleagues made five anchor stations across the Straits of Florida before leaving the ship in Miami. The scientific work was finished then, and *Atlantis* sailed straight for Woods Hole. The air and the sea grew cooler, the sky was more often overcast, and on May 7 in a thick fog *Atlantis* moved hesitantly into Great Harbor.

"Intimate secrets in the marital affairs of the deep-sea devilfish . . . were revealed here yesterday afternoon as the Oceanographic research ketch *Atlantis* returned to her base after more than four months of scientific research work in southern waters," the *New Bedford Standard Times* reported on its front page.

As planned, cruise 65 was the last of the Oceanographic Institution–Bingham Foundation cruises. In spite of their many and varied difficulties, the program had been a success. The most impressive contribution was Parr's hydrographic survey, for which he used the hundreds of temperature and salinity measurements taken from *Atlantis* to compute the general stratification and circulation of the Central American seas. In simple terms, he had found that the main flow of water, both at the surface and at intermediate depths, came into the Caribbean from the southeast and moved north and west in an undulating pattern that refected the great peninsulas of Honduras and Yucatan. The current then entered the Gulf of Mexico via the Yucatan Channel, swept around the gulf in a clockwise direction, and reentered the Atlantic through the Straits of Florida, where it became known as the Florida Current. Although a few major changes and many refinements have been made in this picture over the past forty years, physical oceanographers still read Parr's papers and consider his work the foundation of their own.

In Parr's words, the cooperative cruises also undertook "the first absolutely quantitative study" of sargasso weed. After mapping its distribution and estimating that some seven million tons of the weed are floating around in the Sargasso and Central American seas at any one time, Parr confirmed the hypothesis that *Sargassum* must be able to grow and reproduce at sea. There was far too much of it, he said, to have come exclusively from rocky beds near shore.

Progress was also made in straightening out the taxonomic confusion of the flying fish family. It was discovered, for example, that juvenile fish are sometimes very different in form from adults of the same species. Before this fact was recognized, the two forms were often given different names and considered to be separate species. The food of the flying fish was found to be largely copepods, a form of animal plankton. Of course the flight of the flying fish was carefully observed also. It was described as a gliding rather than a true flying action, although Parr's assistant did not expect many to take his word for it.

"It seems quite hopeless," he wrote, "to convince the casual observer that their flight is not impelled by some form of wing flapping."

All in all, the four cooperative cruises to the Caribbean made a considerable contribution to scientific knowledge. A lot more was known about the hydrography and certain aspects of the biology of the area in 1937 than had been known before. And a great deal was being learned about running a research vessel. *Atlantis* was still a young ship, and with the experience being gained by her scientists and crew on every voyage, it seemed that with careful planning she could undertake almost any oceanographic project imaginable.

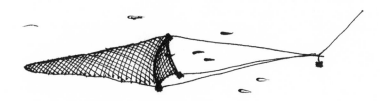

4 Northern Voyages

Most human history has not afforded men much chance to pursue their curiosity, except as a hobby of the rich or within the refuge of a monastery somewhere. We can count ourselves fortunate to live in a society and at a time when we are actually paid to explore the universe.
— Henry Stommel, on receiving the Bigelow Medal, 1974

By the end of 1935 it could fairly be said that the foundations for Henry Bigelow's program of basic oceanographic studies had been laid. Scientists at the Institution had published some ninety papers that ranged in subject from the circulation of the western North Atlantic and the hydrography of the Caribbean to the daily migrations of plankton, the seasonal cycle of chemical nutrients, the bacterial populations of marine sediments, and the discovery of undersea canyons. Each Woods Hole biologist, chemist, geologist, physicist, and meteorologist had undertaken some investigation and not a single discipline had failed to make a contribution. Yet Bigelow was not entirely happy with the program. In his view it lacked a truly interdisciplinary project, and he believed that only an investigation of this sort, carried out in all the seasons of several years on a relatively small, self-contained body of water, could eventually produce an accurate picture of the ceaseless cyclic processes of the ocean.

Before becoming director of the Institution, Bigelow himself had undertaken a study of this kind in the Gulf of Maine. From 1912 to 1927 he had made frequent cruises over this roughly rectangular body of water that lies between Cape Cod and the southeastern arm of Nova Scotia, and he had gathered materials that were later incorporated into three admirably complete monographs on the fish, plankton, and physical oceanography of the area. It was often said that Bigelow's work had made the Gulf of Maine the best-known body of water in the world, oceanographically speaking. Yet he was the first to point out that much more could be learned if his solitary efforts on a small fishing schooner were supplemented by those of several scientists on the modern *Atlantis*.

The difficulties in attaining this goal were immediately apparent. *Atlantis* might not be powerful enough to breast the frigid gales that sweep across the Gulf of Maine in winter, and the mechanics of bringing together a half-dozen busy, independent scientists during the academic year when they were scattered far from Woods Hole might be too complicated. Considering these problems, Bigelow and Columbus Iselin had first decided that in wintertime *Atlantis* could be used more profitably in southern waters, and so, although the ketch had been sent into the Gulf of Maine a dozen times in warm weather, when winter came, off she went to the Caribbean with Albert Parr and others from the Bingham Oceanographic Foundation. Late in 1935, however, Parr had postponed his winter cruise for a year and *Atlantis* was left without plans. With the Institution's own scientists already committed to their universities, it seemed that Iselin had no choice but to leave the ship tied to the dock in Woods Hole all through the winter snows and cold spring rains.

But Iselin began to consider an uncommon scheme. Could he, he wondered, send the vessel on an extended hydrographic cruise to northern waters overseen by a single scientist and his assistant? Such a voyage would give him a chance to see how the northern portion of the Gulf Stream behaved during the winter and to study the forces that prevent the icy coastal waters from moving offshore.

Iselin discussed the plan with Alfred Woodcock, the assistant he had in mind, and the two agreed that although success would depend largely on the weather, and although the weather in the North Atlantic in midwinter is notoriously bad, it was worth a try.

Atlantis was thus prepared for sea that February, and emergency rations were added to her stores in case a storm drove the ship eastward past Halifax or her engine broke down where there were no facilities for repair. When Iselin's wife became ill and he relinquished his plans to head the expedition, still the preparations went forward.

"You will undoubtedly have trouble at this time of year," wrote Iselin in the sailing directions. There would be storms, fog, rain, sleet. "If field ice is reported along the southwest edge of the Grand Banks, the shallow stations may be omitted."

Early on the morning of February 26, *Atlantis* eased away from her berth and steamed into the calm waters of Vineyard Sound. The sea was a brilliant blue that winter morning, and the ketch seemed scarcely to disturb its stillness as she grumbled up the sound. It was a slow and peaceful start.

Two days later the real work of the cruise began. Under all sail *Atlantis* moved briskly along to the south-southeast and at 4:50 A.M. in total darkness hove to on the first station. The objective was to run a series of long northwest–southeast zigzags along the eastern edge of the Gulf of Maine between the shallow, icy waters of both Georges and Grand banks and the deeper, warmer offshore water beyond the Continental Shelf. The first line of stations began, therefore, in raw winter weather on Georges Bank, where the men clumped along the deck in seaboots and heavy jackets handling cold metal water bottles or trimming a sail with stiff, half-frozen lines. The transect ended offshore, where the air temperature rose into the low sixties and the water was so warm that dolphin swam in the ship's bow wave and Portuguese men-of-war could be seen sailing off under their irides-cent floats. Under either condition, mild or miserable, Woodcock appeared in the deck lab for every station. In the highly unusual role of single scientist he felt himself responsible for all data gathered. As if his scientific responsibilities were not enough, he chose to spend hours standing watch. Huddled in the bow or clinging halfway up the mast, he watched all that moved through sea and air.

Woodcock was a lithe young man of slender build whose interests had been turned from experimental farming to marine science by a series of accidental encounters. Raised in Atlanta, Georgia, he had attended an agricultural school in the Northeast, and part of each year had worked in a fruit orchard. The orchard's owner had a yawl, and when work was light, he, his family, and Woodcock would take a few weeks off and go cruising along the New England coast. On one such trip they had put in at Woods Hole, and while Woodcock was getting his hair cut he overheard the barber talking about the new Oceanographic Institution and the beautiful ketch that was then on the ways at Copenhagen. On an impulse, he had gone to see Iselin.

"I liked him at once," Iselin recalled, "and signed him on [for the maiden voyage] as an ordinary seaman at $45 a month, feeling that he would be a steadying influence among what looked like a rather turbulent crew. . . . From the outset it was obvious that Woodcock was much more than a young sailor. . . . He has been scientifi-cally . . . productive. . . . In fact, he is a remarkable person."

On the maiden voyage Iselin had encouraged Woodcock to keep records of sea conditions and weather, and this dual role of sailor–lab technician, which meant that Woodcock was forever interrupting his holystoning to read a thermometer, had been resented by Clowser. Yet Woodcock had outlasted Clowser, had given up sailoring to be a full-time technician, and by the winter of 1936 had spent more time aboard *Atlantis* than any of the other scientific personnel.

In the upper laboratory on *Atlantis* Al Woodcock detaches the net from a Clarke-Bumpus plankton sampler. *(George L. Clarke photo.)*

On this cruise, as on all others, Woodcock indulged his appetite for observation, and as the ship worked north, he watched sleek black-and-white auks dive on the hull to feed on the organisms that grew there. He watched the gulls soar, and when the water was warmer than the air and updrafts were created, he could see the birds circling higher and higher without a flap of wings. Sometimes young, dark-feathered herring gulls, gray-and-white adults, and old black-backed gulls with tattered feathers and long, hooked beaks all spiraled upward together in the same updraft. Under other conditions they formed what looked like flat sheets of birds that moved like regimented gliders against the wind.

During windier days, and there were plenty of them that winter, Woodcock watched the gulls swoop and dive in accordance with invisible currents of air that were deflected by the waves and by the sails of the ketch. He noticed that when gusts exceeded 30 miles per

A Nansen bottle with its reversing thermometers is attached to the hydrographic wire in northern waters. *(Courtesy Woods Hole Oceanographic Institution.)*

hour the gulls' wings began to flutter, and they flew off with their inexplicable weather sense to calmer regions. The shearwaters, however, could still maneuver, and Woodcock loved to watch them make daring swoops to within inches of the hissing waves, then veer into a trough, where for a moment they were protected from the wind. The shearwaters could fly and feed until the winds exceeded 40 miles per hour; then, unlike the gulls, they rode out the storm resting on the waves.

In early March *Atlantis* put in to Halifax and tied up for reprovisioning at the Western Union dock. "Woodcock has been troubled with eye inflammation," McMurray wrote, "due to a combination of eyestrain and cold."

Dense fogs and sudden storms kept the ship in Halifax (with its strict liquor laws) for a full week, but on March 14 a pilot was taken aboard and the ship steamed out of the harbor.

After a week of routine sampling with only occasional interruptions from fog and wind, the ketch encountered the troubles Iselin had predicted. First Officer Kelley's weather maps, drawn up each noon from information received over the radio, showed a bad storm making up to the south, and *Atlantis* hove to to avoid it. The wind freshened, blowing long streamers of fog through the shrouds, and McMurray ordered the main trysail bent on in anticipation of a gale. Hour after hour the ship seesawed to windward. Heavy rain squalls and a leaden sky made it impossible to take a sight on sun or stars, and within a day the ship's position in relation to the storm could only be surmised. Still McMurray kept the ship hove to for a second day. The wind had dropped, but a majestic swell was running, and the rain came pouring down. Sheets of water drummed against the flat trysail, sluiced off its foot, splashed on decks and hatches, and streamed through the scuppers. Rain fell upon the sea with such intensity that the crest and peak of every wave were beaten down and only long, rounded swells rolled past the ship like undulating sheets of watered silk. Over the radio came news that a Dutch ship, caught in the storm to the south, had lost her rudder and might need assistance from *Atlantis*. Luckily she did not.

On the third night a few holes opened up among the clouds and Kelley considered himself "damned lucky to get one pop at a star before the sextant was covered with water." A heavy sea still rolled in from the southwest and the wind began to freshen. Coming up on deck, one of the mates noticed that the lashing fastening one of the brass slides to the trysail was beginning to part and sent Ernest Siversen, a fifty-four-year-old sailor of uncommon ability, up the mast to repair it. Siversen went up to the halyard blocks, some fifteen feet above the deck, and as he tried to replace the chafed roband, the ship made a great roll and Siversen fell to the deck. He was up in a moment, cursing mightily. Kelley treated him for a cut above his left elbow and pain in the shoulder, and *Atlantis* moved on toward her next station.

The storm to the south had finally blown out to sea, but as *Atlantis* moved into its place she was struck by a gale from the northwest. For two and a half more days the ship lay hove to, bucking and pitching over the rough gray sea, and when the wind went down, the fog closed in. Whether a man was standing at the wheel, posted in the bow as lookout, lying in the fo'c'sle, or sitting in the saloon, the resonant complaint of the ship's foghorn could be heard each minute of the day and night. A single groan meant that McMurray had put her on a starboard tack; two, a port tack; three, the wind was abaft the beam and the ship was running free.

On March 28 the weather cleared, the crew did some work on deck, and the weather map showed a gale coming in from the southeast. It arrived at ten that night. This one passed as quickly as it had come, and by the next morning *Atlantis* could carry all sail and go tearing back toward Georges Bank to start another line of stations. By midafternoon she had arrived at Meade Shoal, where the line was to begin and where the seas were extremely steep and short. It was questionable whether a station should be attempted under these conditions, but Tom Kelley decided to try. While the ship's foghorn repeated monotonously one long and two short — unable to maneuver — Kelley positioned himself on the short platform that projected outboard near the hydrographic wire. With every other roll of the ship he was immersed to the waist in swirling water. He was not tethered to the ship with a lifeline (he did not favor such constraints) but used the line that served as the platform's railing as both support and guide. He backed up against it, pressing the line into the small of his back, and if, when the ship dumped him into a confusion of icy water, the railing fell to his hips, he knew he was about to be washed off the platform. It was, he said later, the worst station he had ever worked.

At eight that evening a lifeboat chock was carried away and shortly afterward the dreary rain turned to sleet and then snow. By morning a gale drove the snow horizontally across the deck and McMurray ordered the ship hove to for another day and a half in water so cold and rough that it resembled the ridges of an Arctic ice field. By this time the trysail had been up so much of the time under such enormous strain that it was beginning to pull its track off the mast. New screws were put in without mishap and for the remaining eight days of the voyage the ship continued to alternate between running lines of stations and heaving to in thick weather.

Before dawn on April 8 the flashing light of the gas buoy off Gloucester was picked up, and Siversen, whose shoulder still gave him a great deal of pain, was put ashore with an escort to steer him past bars and into a hospital. There it was discovered that the shoulder was broken. Siversen remained behind when the ship got under way again and headed for Woods Hole. She came through Pollock Rip in the dead of night, slid past the lights of a few early risers in Vineyard Haven, and was fast to the dock by 6:00 A.M. After five stormy weeks at sea the crew knocked off for breakfast and returned at 8:00 A.M. to unload the water samples that had been collected in spite of the gales.

Later that spring, when the weather had moderated, Iselin went

Working in the Gulf of Maine, *Atlantis* would occasionally get so thoroughly covered with ice that her bow would be down. The remedy was usually a run toward the Gulf Stream, but sometimes the crew would have to chip and sand. *(Courtesy Woods Hole Oceanographic Institution.)*

out with Woodcock for a three-week voyage to the same area, and a third cruise was made in the fall. On that trip Woodcock made a cast in water about 2,200 fathoms deep and hit something at 130 fathoms. The cable went slack. It was drawn up, let down again, and this time dropped the full 2,200 fathoms without interruption. Woodcock was baffled and could only surmise that he had first dropped the water bottles on the back of a whale.

Iselin was extremely pleased with the temperature readings and water samples that had been taken in this difficult region of the sea. Even in the early years of oceanographic exploration it was almost unheard of to send a vessel out with a single scientist, and he gave Woodcock much of the credit for the eventual publication of two of his own major papers, "The Development of Our Conception of the Gulf Stream" and "The Study of the Circulation of the North Atlantic."

Woodcock's temperature and salinity measurements had certainly shown that work could be done in parts of the Gulf of Maine in the severest weather, and Siverson's shoulder and the frequent need to heave to had shown with equal clarity the price that had to be paid for the data. Consequently, plans for a year-round program in the gulf were again held in abeyance while *Atlantis* was used in the Caribbean by Parr in the winter of 1937 and by a group from Harvard and the University of Havana in 1938 and 1939.

But in the fall of 1939 a more organized attempt was made to launch a comprehensive study of Georges Bank, the famous fishing grounds in the southeastern portion of the Gulf of Maine. One of several reasons for undertaking the project was a desire to study the productivity of this rich area, but an equally important goal was to pin down the elusive state of true interdisciplinary cooperation that Bigelow considered so important.

In the nine years that Bigelow had directed the Institution he had had little trouble in getting biologists, chemists, physicists, and geologists to talk together — he encouraged them (with free ice cream) to attend a Thursday-evening seminar that, he said, "gave 'em a chance to sniff each other" — but he had not been able to get them to adopt a common project and really work on it together. The cruises to Georges Bank were organized with this end in mind.

Again there were problems. In 1939 the professional staff at the Oceanographic (excluding visitors who came for short periods) had grown to include about twenty-four persons, of whom perhaps eight had sufficient knowledge and experience to organize part of the project. There was, for example, only one geologist, Henry Stetson. In

1939 Stetson was too busy to start mapping sediments on Georges Bank, and Bigelow was too much of an individualist to pressure him into joining the project. The same problem arose with physical oceanographers — Iselin didn't have time, and the only other one affiliated with the Institution lived in Canada and would not be available for cruises in the fall, winter, or spring. Chemistry was better represented, but the Institution's chemists were still struggling to perfect methods of analyzing nutrient chemicals such as phosphorus and were not really ready to run the tremendous number of routine analyses that were necessary.

As was usual in oceanographic laboratories in the prewar years, Woods Hole's biologists nearly equaled in number all the other kinds of scientists at the Institution combined. It was not surprising, therefore, that the work on Georges Bank was largely done by George Clarke, a zoologist, Gordon Riley, a marine physiologist from the Bingham Oceanographic Foundation, and two laboratory technicians from the Oceanographic — Dean Bumpus for plankton and Dayton Carritt for chemical work. Clarke, with his knowledge of plankton and light penetration, outlined a foresighted program of quantitative measurements which, if made on Georges Bank for three years in a row, and at all seasons, might point to the factors that enabled the area to support such a rich population of fish.

Between September 1939 and June 1941, when the project was prematurely ended by the war, *Atlantis* made eleven trips to Georges Bank, and although the seasons changed and the weather varied, a standard program of observations was carried out on each cruise. Typically, a voyage lasted about ten days (longer in the winter, when the weather was bad), and during that time sampling and towing were carried out at stations that together formed a grid covering the bank.

On cruise 113, for example, *Atlantis* left Woods Hole early one April afternoon and under a light breeze steamed out of Great Harbor and began the familiar journey up Vineyard and Nantucket sounds. Under engine power alone, the ship crept along past Vineyard Haven, Hedge Fence, and the other well-known although largely invisible seamarks. Toward evening the men gathered in the saloon for the first and last unhurried, uninterrupted meal the scientists would have together until the work of the cruise was finished.

Dean Bumpus was chief scientist, a title with more responsibility than honor, and it was up to him to see that the work proceeded smoothly. "Bump" had first come to the Institution in the spring of

1937, and, having found the work to his liking, had left graduate school to follow Clarke and his plankton to Harvard. There, during the winters, he counted, measured, and identified the plankton he had caught from *Atlantis* the previous summer. By 1941 he had become an oceanographer by Bigelow's "catch 'em young and rear 'em" method and was responsible not only for much of the work done at sea but also for finding extra help to fill out the scientists' watches. Claiming that he was always shorthanded, he began signing on "blueberry pickers, ski bums, and hungry artists, most of them disinterested," who would go to sea in exchange for room and board. (Relatives were also pressed into service. Clarke's brother-in-law made his first cruise as a teenager, and Bump once signed on his own father.)

On cruise 113 the extras outnumbered the regulars. With Bumpus and Gordon Riley were Val Worthington, who had just flunked out of Princeton and was making his first voyage; Pete Davidson, the brother of a woman who worked at the Oceanographic; and Fritz Fuglister, an artist and violinist who had just finished painting a mural on books and bookmaking for the WPA. These five men sat in the saloon after supper and, with their charts and notebooks spread over the table, decided how to divide the work. That chore finished, they had a smoke and then turned in early to catch some sleep.

The ship rolled on without interruption until 4:10 A.M., when the first station was begun. As *Atlantis* slowed down and hove to in a flat calm, her deck lights flashed on and from somewhere deep in her hull came the familiar *cheep-cheep-cheep* of the sonic sounder.

"Forty-nine fathoms," called the man at the Fathometer.

"Forty-nine fathoms," repeated someone in the upper lab, who had waited for that information before deciding how to place the water bottles along the hydrographic wire, and "Forty-nine fathoms," echoed the mate in the chartroom as he entered the figure in the log.

There was a curious and rather pleasant sense of isolation on deck that night, as if the active portion of the world had contracted within the ring of deck lights. Nothing could be seen beyond particles of fine mist that danced in the lights or heard above the thrum of the engine and a gentle gurgle of water passing along the hull. The ritual of the water bottles ensued, and within the hour they were passed across the deck, let down into the black water, and brought back to their racks in the lab. Then, as the sky began to lighten and the world to expand, plankton nets were broken out of a box on deck and brought to the rail.

Dean Bumpus with the chariot, a big net designed to roll along the ocean floor. *(Herb Gardner photo.)*

First to go over the side were three of the newly invented Clarke-Bumpus plankton samplers. Each had an opening and closing mechanism so that it could be made to sample plankton from one specific layer of water only and for a precise number of minutes. The sampler was also equipped with a flow meter that registered the exact amount of water that passed through. (The objective in using the sampler was to estimate the amounts as well as kinds of plankton that lived on Georges Bank, even though McMurray maintained that to do so "you'd have to strain the whole goddamned Gulf of Maine.")

Once the samplers had trapped their microscopic prey, a stramin "chariot" was put over to sample fish eggs and larger forms of animal plankton. The chariot was a large net, a meter and a half across, mounted within an outer hoop that had two small wheels on its bottom edge. Unlike a conventional net, it could be let down all the way to the bottom, where its wheels prevented the net from digging into the sediments and where, as had been hoped, it sampled animals that lived just above the bottom of the sea floor.

By 7:00 A.M. the chariot was back aboard and *Atlantis* got up speed and headed south by east toward the next station. Smells of bacon and toast came up on deck, and the scientists, who were working six hours on and six off instead of the usual four on and eight off, hurried to the saloon for breakfast before the second station began at 8:30. This one was finished by 11:00 A.M., and had there been sufficient sunlight the ship would have hove to at noon to make light-intensity measurements. A third station was made that afternoon as patches of fog drifted past the ketch, a fourth at 9:00 P.M. as the wind freshened and tugged at the big net when it went over the side, and a fifth at 2:41 the next morning.

On the sixth station Ernest Siverson, who was filling in as third mate, operated the trawl winch, and, because he lacked experience (and perhaps because he was nearsighted and slightly deaf), he neither saw nor heard the chariot rise from the water and pass above the rail. Suddenly there was a crunch of metal on metal and the net's bridle started in through the blocks at the end of the trawling boom. The gear had been two-blocked, and predictably the wire broke and the net dropped back into the water and disappeared. A second net was immediately rigged from extra parts and the stations across the bank continued with hardly a pause. The next night, however, when a gentle southwest wind blew fog across the ship and the same mate was at the winch lowering the chariot, he suddenly found that he had no control of the wire.

"It's running free!" he shouted, waggling the winch lever back and forth with no effect. Someone moved forward to grab the wire but was stopped by Fuglister, who knew that the rough, frayed cable would shred his hands. Helplessly they watched the cable snake across the deck. A broken end came up out of the hold, and the net and some hundred feet of wire disappeared over the side. Apparently the first accident had weakened the cable, and it had broken as it was being reeled off the big drum in the lower hold.

End of towing, thought Fuglister with satisfaction, that portion of the work always having seemed monotonous to him, but when Bumpus was roused from his bunk at 1:00 A.M. and told of the second accident, he began at once to assemble spare parts on deck. He had no trouble finding a net, but a hoop and bridle had to be constructed and there were no more "fish," the heavy, streamlined weights that kept the end of the wire from angling up toward the surface as the nets were towed. Bumpus persisted, however, and as the stars faded and the sky and sea turned gray, he produced, from some deep recess in the ship, two heavy steel balls.

"Obscene!" shouted Fuglister with delight as Bumpus fastened the balls together and posed astride them.

The jury-rigged net was let down to the bottom, towed, and retrieved successfully. As usual, the deck wash was turned on as the net came to the surface, and once the bridle and hoop were hoisted inboard the net itself was sluiced down. Still running, the hose was stuck down a scupper while the cod end of the net was brought in. As Bumpus and Fuglister congratulated each other on their makeshift net and dumped the catch into a tub, the deck wash worked free and, like a giant garden hose gone wild, writhed along the deck shooting cold salt water at the two men. The water seemed to know where Bump was, and as he dodged back and forth across the deck, shouting and spluttering, Fuglister tried to keep behind him.

"Christ, it sounded like a flock of parrots being murdered," recalled one of the sailors.

The deck wash brought under control, cruise 113 continued without storms or accidents; for a trip to Georges Bank, the voyage was unusually peaceful. More often spring was a cold, unsettled time there, and the work on *Atlantis* was often hampered by roisterous winds that blew across water still chilled by winter.

In summer, under a warm sun, cruises to Georges Bank were exuberant. The prevailing southwest wind sometimes blew so steadily across the sparkling, white-capped waves that *Atlantis* could reach across the banks at fourteen knots. She made a fine hissing sound then and her white wake flew out behind. In autumn the weather could be even more beautiful, and on those last warm days the sky, the sea, and even the ship herself seemed to give back the warmth they had gathered in all summer.

Inevitably, winter came, and the cruises became a misery. The sea was rougher than a cob, Bump used to say, and on the coldest days he remembered waves washing in through the scuppers and leaving a film of ice on the decks, so slippery that sand had to be strewn about.

The last of the Georges Bank cruises before the series was cut short by the war, was made in June of 1941, and papers describing the experiment came out in the mid-1940s. In several Clarke described the flow of energy that coursed through the ecosystem on the banks. The ultimate source of energy was, of course, sunlight, and this energy was captured by plant cells that in turn were eaten by animal plankton. Small fish fed on the plankton and larger fish preyed upon the smaller fish. There were also many cod and haddock on the bank, and these fish, largely bottom feeders as adults, fed on crabs, worms, and mollusks that either ate smaller members of the sea-floor community or survived on the remains of plankton that drifted down from the surface waters. Bacteria helped complete the cycle by attacking the dead and discarded, which in the process of decay returned to the water those nutrient chemicals needed by plant cells to keep the cycle turning.

A few years later, Gordon Riley, Dean Bumpus, and a young astronomer-turned-physical-oceanographer, Henry Stommel, tried to fit numbers to the components of this dynamic budget and in so doing created one of the first theoretical models of a balanced marine food chain. It was a fine piece of interdisciplinary work, and if the war had not intervened, still more scientists of still other disciplines would have been added to the team, as Bigelow has hoped.

5 Submarine Canyons and Continental Shelves

I don't know of any other human occupation, even including what I
have seen of art, in which the people engaged in it are so caught up,
so totally preoccupied, so driven beyond their strength and
resources. . . .
  Scientists . . . are, despite their efforts at dignity, rather like
young animals engaged in savage play. When they are near to an
answer their hair stands on end, they sweat, they are awash in their
own adrenalin. To grab the answer, and grab it first, is for them a
more powerful drive than feeding or breeding or protecting
themselves from the elements.

— Lewis Thomas, *The Lives of a Cell*

Rough, tough, and handsome, Maurice Ewing jumped from
the driver's seat of his old Ford, Fluzybell, and without a stretch or a
yawn began unloading spools of wire, spare parts, toolboxes, and
crates of half-completed equipment onto the Institution's pier. Doc,
as he was often called, and his students from Lehigh University had
driven nonstop from Bethlehem, Pennsylvania, to Woods Hole, and
now, in what appeared to the citizens of that quiet community to be a
burst of frantic activity, they were preparing to go out on *Atlantis*. It
was October 1935, and Ewing, rebounding from the disappointment
of a totally unproductive cruise aboard a Coast and Geodetic Survey
vessel, was intent upon studying the structure of the sea floor with a
novel combination of dynamite and seismographs.

The young geophysicist from Texas had first become interested in
the structure of the earth when he had gone prospecting for oil in the
shallow bayous of Louisiana during his college vacations. After
graduating, and upon accepting a teaching position at Lehigh, he had
considered trying at sea essentially the same techniques he had used
in the bayous. His intention was not to discover oil, but to study the
structure of the continental shelf. He believed that if he set off ex-

plosives on the sea floor and used a seismograph (geophone) to record the sounds that traveled back to the ship through the water, the mud, and even the rocks that might underlie it, it would be possible to determine the thickness of the sediments and learn something of the geologic structures that lie beneath them.

With these objectives, Ewing and several of his students and supporters hurried to Woods Hole and joined Columbus Iselin and the Institution's sole geologist, Henry Stetson, aboard *Atlantis* for a short test cruise.

Shortly after breakfast on the sunny and pleasantly mild morning of October 17, the ketch cast off her lines and steamed out into Vineyard Sound. As she cleared Nonamesset and turned down the sound, Iselin called for a council of war so that Captain McMurray and the mates might know what kind of work was planned. Ewing's two students came aft from the deck lab, where they had been stacking boxes of dynamite, and Doc emerged from his cabin, where he had been stowing detonators under his bunk. All were dressed in dirty sweatshirts and grease-stained pants, and all wore the distracted expression that the crew had learned to associate with impossible scientific schemes.

As the men stood leaning against the wheelhouse watching the Elizabeth Islands slide by, Ewing earnestly described his plan. He wanted to attach a geophone, encased in a heavy pipe, to the ship's anchor cable and lower it to the sea floor. While the phone rested on the bottom, one of the ketch's whaleboats would sail away, firing bundles of dynamite at predetermined intervals.

McMurray did not like the idea at all, but with Iselin aboard, the scheme would at least be given a try.

By early afternoon *Atlantis* was riding over some fifteen fathoms of water just south of the small island of No Mans Land and two crewmen were instructed to bend a kedge onto the trawl wire, which had been led forward and out through one of the hawse pipes. Once the anchor was in the water, Ewing broke out the geophone. When it was attached to the cable, the trawl wire was paid out until both anchor and instruments rested on the bottom. (The regular anchor and chain were not used because they could be wound in only by hand.) At the same time Ewing's students had been filling the uppermost whaleboat with explosives, radio equipment, and other necessities, and at 1:15 P.M. the order was given to lower away.

"Thus I found myself in the whaleboat loaded deeply with boxes of TNT [actually dynamite]," wrote Iselin, with John Lindstrom at the tiller and a student of Maurice's named Albert Crary. . . . His job was to prepare and fire the charges. There was a moderate sea running and my first worry was that Crary would become sick. [Soon Crary did begin leaning over the side and spitting as he rigged the charges. He was chewing tobacco.]

We had a small radio in the boat which we managed to keep reasonably dry. We also had a six-foot pole with a one-foot-square plank nailed to its end. The blasting cap attached to the outside of this plank interrupted the radio signal when fired and also fired another cap in the submerged charge. Thus *Atlantis* recorded the instant of firing and then the arrival times of the signal.

All afternoon the whaleboat sailed back and forth within half a mile of *Atlantis*. The men on the ketch could see her lugsail trimmed

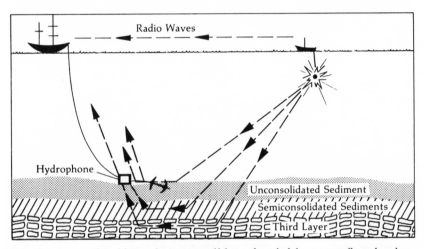

Sound waves produced by explosives set off from the whaleboat are reflected and refracted at the ocean floor. Reflected waves bounce off the bottom like a signal from an echo sounder. Refracted waves travel through the sediments. Since their travel times depend on the types of sediment they pass through, the refracted waves give oceanographers an idea of the thickness and composition of subbottom layers. On later cruises the hydrophone was suspended at the surface beneath the receiving ship.

to catch the rising southwest wind, and on a reach she fairly flew over the glittering waves. Then down went her sail as a shot was prepared, and, looking much smaller, she bobbed aimlessly over the swells. A low shudder passed through *Atlantis* as the dynamite went off. Up went the lugsail again and the whaleboat skipped off to repeat the procedure.

As the sun set and the cook ungraciously agreed to keep dinner waiting, a flag was run up to signal the whaleboat to come in.

After 1935 the whaleboats on *Atlantis* were routinely used for seismic refraction work. The boats were built by Phoebe Beetle at her family's boatyard. Phoebe was six feet tall, had iron-gray hair, and spat tobacco juice. The boats she built "were right." *(Don Fay photo. Courtesy Woods Hole Oceanographic Institution.)*

Since the small boat could not be sent out after dark, Iselin imagined that Ewing would stop working for the night, but he did not. Ewing had ingeniously figured a way of floating charges off the stern of the ship on blocks of wood. These floating bombs were tethered loosely by a thin wire connected to an electric detonator and when sufficiently far astern could be fired at will. The geophone was still on the cable out ahead of the ship and was still recording any vibrations it received.

Charge after charge floated away from the stern. After the watch changed at midnight, only Ewing and Iselin continued working on the quiet ship. While Iselin was "tending to the shooting and I was tending to the recording," wrote Ewing, "we got the wire to one of the charges . . . caught somewhere under the stern in the rudder or the propeller."

Over coffee in the empty saloon Ewing and Iselin tried to estimate the probability of an explosion once the propeller was started and the extent of the damage if an explosion should occur. The best move, it seemed, would be to start the propeller dead slow, then speed it up gradually. Anger and righteous indignation battled in McMurray when he was told of the predicament, but he moved *Atlantis* ahead in the dead of night, and there was no explosion. Apparently the wire had broken, dropping the dynamite harmlessly to the bottom.

Ewing disappeared for several hours' sleep but was up before dawn to prepare the whaleboat. Despite a rising swell that vigorously lifted and dropped the whaleboat alongside the ship and quickly obscured the small boat once it had pulled away, tests were made throughout the morning and afternoon. The small boat was sent a mile and then a mile and a half away before returning to *Atlantis*.

As the sky clouded over and evening fell, the ketch started in for Woods Hole. She passed through a fleet of fishing boats from Boston and New Bedford and shortly before midnight tied up at the Institution's pier. No useful records had been procured on this test cruise (probably because the geophone had been dragging along the bottom on the cable), but Iselin believed Ewing to be on the right track, and he advised Bigelow to give him more time on *Atlantis*. Consequently the vessel was off again within the week with Ewing prepared to make a series of seismic measurements from Cape Henry, Virginia, to the edge of the continental shelf, some eighty miles to the east. On this cruise Ewing managed to get records from four refraction lines, and from these records came the first glimpse of the shelf's structure. According to these data, the shelf was com-

Maurice Ewing rarely slept more than two or three hours at a time during a cruise. *(Don Fay photo.)*

posed of three layers. The first was of loose sediments several thousand feet thick; the second was a deeper, thicker layer of semiconsolidated sediments; and the third was of a still denser material that Ewing surmised was bedrock. Because the top two layers together were 12,000 feet thick, similar to the arrangement of sediments on land, Ewing correctly assumed this portion of the sea floor to be continental in character. The arrangement of sediments and bedrock farther out to sea was largely a matter of speculation.

Although Ewing was most anxious to pursue his seismic work, believing it to be by far the most important project with which he had yet been connected, he was unable to come to Woods Hole again until the summer of 1937. In the meantime, there was another study of the sea floor going on at the Oceanographic, one as different in character as the project's director, Henry Stetson, was from Doc himself.

Henry Stetson, a paleontologist-turned-geologist, was a mild-mannered man and, like many members of the Institution's staff, a well-to-do New Englander and Harvard graduate. Just as his friend and contemporary Columbus Iselin had been introduced to oceanography aboard his own schooners, Stetson had bought himself a Friendship sloop as an undergraduate and had taken water samples and even some short cores for Dr. Bigelow during school vacations. He was in his early thirties when he came to the Oceanographic, and a friend described him as "a modest man who was suspicious of theorizing and had no use for shortcuts or fidgety doo-dad." In short, he was a proper Yankee and very unlike Ewing. Their approach to almost everything differed. Their response to *Atlantis* was no exception. While Ewing tolerated the ketch, making no secret of his preference for fully powered ships that could move quickly from station to station with little regard for the weather, Stetson loved the ship and had in abundance the patience and sensitivity that sailing her demanded.

For most of each year Henry Stetson worked at Harvard's Museum of Comparative Zoology, but in the summers he pursued a project of local sea-floor sampling at Woods Hole. Then in 1934 he became interested in some curious undersea canyons that had been discovered at the edge of Georges Bank.

In the very early 1930s the U.S. Coast and Geodetic Survey had recharted Georges Bank, using sonic sounders for the first time, and had found a large number of steep-sided gorges cut into the southern edge of the bank. These canyons ran back into the continental shelf some five to twelve miles, were themselves two to six miles wide, and had floors that lay from 1,200 to 8,000 feet below the surface of the sea. One of the first geologists to become intrigued by the canyons was Francis Shepard, who had gone out on a Coast and Geodetic Survey cruise himself. By 1934, however, he had chosen to study similar features in southern California, and the eastern canyons were thus scientifically abandoned within a year or two of their discovery. When Stetson decided to take up the problem, he chose three canyons where Shepard had found "an area of such relief and irregularity that it dwarfs by comparison anything above water in eastern North America and must rival the grandest topographic features of the west."

In August of 1934, Stetson equipped *Atlantis* with heavy steel dredges and a simple coring device and sailed off for Georges Bank. A routine was soon developed during those cool summer days whereby the men on *Atlantis* first located a canyon by means of the ship's Fathometer, then marked it with a buoy. The ship was then positioned over the middle of the canyon and a rock dredge let down to the sea floor. The dredge was allowed to stay there as the ketch steamed slowly toward one edge of the canyon while cable was paid out to the dredge at a commensurate speed. Once the ship was about a mile ahead of the dredge, the winch was stopped, and as the vessel proceeded, the dredge began dragging across the canyon floor and bumping up its steep wall. Even before the roughly box-shaped device with its link-and-ring bag was brought back aboard, Stetson could get a fair idea of the kinds of sediment it was encountering by looking at the dynamometer that registered the tension exerted on the cable by the dredge. A strain of about 3,000 pounds suggested muds and oozes, 7,000 to 10,000 pounds indicated older, semiconsolidated sediments, and short, alarming periods of much greater strain that threatened to break the cable told him that a piece of rock had been encountered and cracked off the canyon wall. After the dredge had been dragged up over the lip of the canyon, it was raised, brought aboard the ship, and dumped into big, boxlike sieves. Out of the heavy link-and-ring bag came several hundred pounds of mud, green sand, pebbles, and lesser amounts of cinders, broken rock, and brittle stars. At least once "pink worms on sticks" appeared in a haul, as did "queer pink coral-like growths." Eleven samples of rock were cracked off the canyon wall, and when the microscopic fossils in them were identified and dated, Stetson found the rock to have been formed some 30 million years ago. Geologically speaking, the canyons were relatively young features.

A second clue to the origin of the canyons came from short cores that had been taken from the canyon floors. Using a coring tube that was driven several feet into the sediment by a weight attached to its upper end, Stetson collected twenty cores, and some showed two distinct layers of sediments. Although this was by no means the first time such layering had been seen, it was a fascinating discovery, for by studying the kinds of minerals and plankton that made up each layer, Stetson could reconstruct the approximate conditions under which the sediments had been laid down. The upper layer was composed of recent sediments laid down during a temperate climate, whereas the lower layer contained the remains of cold-water animals that would have lived in an Arctic climate. It seemed logical for Stetson to hope that longer cores would give him a whole suite of sediments, and that these layers, like chapters in a history book, would tell of ice ages, warm periods, and other major climatic events that had helped create canyons, banks, trenches, and even the continental shelf itself.

Stetson returned from his six-day cruise determined to find a better coring device, and almost exactly one year later he persuaded the cautious Bigelow to let him use a new explosive gun corer designed by Dr. Charles Piggot at the Carnegie Institution of Washington, D.C. Onto the ship, then, went the corer, quantities of black powder, and the corer's inventor, Charles Piggot.

A genial man of medium height and middle age, Piggot had been interested for several years in the distribution of radium in marine sediments, and to obtain this element from deeper sediments he had designed and built a coring device that shot a ten-foot steel pipe into the floor of the sea. Piggot's first corer was a totally impractical machine that for every coring operation required seventeen explosions and the precise movement of seventy-five working parts. His second was seventeen times simpler and worked.

Piggot had first used his corer from lighthouse tenders, where, he said, he had been courteously shielded from the hot sun by awnings and handed a mint julep at the end of the day. Life would not be quite so luxurious aboard *Atlantis*, or so he began to suspect as he watched the deck fill with a confusing quantity of equipment and saw sackloads of potatoes roughly dumped into bins, his own duffel heaved down the after companionway, and the scientists, dressed in their oldest clothes, pitching in to load the ship. Earlier he had seen a veritable conga line of sweating, bare-chested sailors come staggering along the pier carrying a furled mainsail onto the ship.

On the afternoon of August 6, *Atlantis* moved through a lively fleet of day sailers, skirted fishermen working for blues off Ram Island, and set sail for a group of submarine canyons off New Jersey and Maryland to give the gun corer its first trials in deep water.

On the seventh, a beautiful summer day, Piggot's corer was assembled on deck and loaded. Macaroni-shaped pieces of gunpowder were packed inside a heavy firing chamber, below which hung the corer's cartridge and firing-pin assembly. Below that was a ten-foot brass-lined core barrel. The entire rig was slowly drawn up from the gently rocking deck and swung over the side. Only then did Piggot, standing on tiptoes, reach over the rail and pull a brass safety ring

Three men climb out on the ship's dredging boom to disentangle the trawl frame that had gotten fouled during a deep anchor station. Under more normal circumstances, one man climbed to the end of the boom and snagged the dredge with a boathook so that it could be pulled aboard. Captain McMurray stands at the rail. *(Courtesy Woods Hole Oceanographic Institution.)*

that allowed the gun to fire when it struck the sea floor (or anything else).

The gun was used four times that day in relatively shallow water just to test its operation. The reverberations of each explosion rattled the ship as the core barrel shot into the hard clay bottom. The barrel, pulled up on deck by an auxiliary cable, required a great deal of pounding and pushing before it would relinquish its stiff cores of gray clay.

Additional cores were to be taken as *Atlantis* moved into deeper water. But during the night the wind freshened, and as the sea made up into a steep chop, the ship began to pitch and roll. There was no question of using the corer then, and besides, many of the scientists were sick.

By August 9 the front had passed, and although *Atlantis* continued to roll and her heavy canvas sails slapped and banged when the ship came on station, the coring was resumed toward evening. This time the device was let down through 185 fathoms of water, and it returned not with the usual two- or three-foot core that could be obtained with a conventional corer, but with a full eight feet of sediments crammed inside the barrel. Stetson and Piggot were so pleased that they reloaded the gun, increasing its charge, and sent it down again.

"First try pulled gear apart," wrote Stetson in the coring log.

During the next four days, through gentle rains and an occasional thunder shower, *Atlantis* worked her way seaward over a submarine canyon. Dredging and coring were done in deeper and deeper water, and by August 13, when the ship turned homeward, the "shotgun" had procured fourteen cores ranging in length from four feet to almost nine feet and taken from water as deep as 1,250 fathoms. The longest core procured by the old corer had been three and one-half feet long.

The new cores aroused a good deal of speculation concerning the creation of submarine canyons. Some scientists suggested that old rivers or waterfalls at the edge of ancient ice sheets had cut the canyons during times of vastly lowered sea level, others favored currents that might cut and scour the continental shelf, and still others thought that the spent energy from "tidal waves" created them.

"The multiplicity of theories that have been put forward is an index of the general perplexity," wrote the patient Stetson, who on a subsequent cruise in 1936 measured the currents that weakly flowed through the canyons and eliminated at least one of the hypotheses.

Stetson continued to visit the canyons once or twice each summer, and although he relied less on the gun corer after an "improved model" went off prematurely against the ship's hull and dented her plates, he kept right on bringing in the sea-floor samples. Stetson was already forming an accurate and fairly detailed picture of the sediments that covered the canyons and the rest of the continental shelf off the eastern and southeastern United States: his work certainly constituted the kind of fundamental knowledge that Bigelow had hoped the Institution would gather.

In the summer of 1937 Maurice Ewing was ready to go to sea again to advance his more ambitious investigations. He had a new objective. Instead of running seismic refraction lines across the continental shelf, he would move into the deep sea and find out how much sediment had accumulated there. If the positions of continents and ocean basins were permanent, as Ewing and most others believed, then he should find sediments many thousands of feet deep lying above the oceanic basement.

Ewing wrote to Columbus Iselin asking for time on *Atlantis* and outlining the new method of shooting that the decision to operate in deep water had forced upon him. Both recording instruments and bombs would be attached at intervals along a steel cable laid out on the sea floor behind the ship. The bombs would then be fired automatically by time clocks.

"Dr. Bigelow has been worrying about the danger of an explosion . . . ," responded Iselin somewhat sheepishly. "This may seem a little fussy to you, but I must lean over backward to convince Dr. Bigelow that we are not being slipshod." And the next day, "We will handle somehow the fifty pounds of TNT which I understand were shipped yesterday. Don't send any more until we have made arrangements with the local authorities."

"I am sorry the explosive arrived there so early as to embarrass you," replied Ewing, "and regret that in addition to the fifty pounds of TNT they have also sent a hundred pounds of SNG."

In all, more than 500 pounds of explosives arrived in Woods Hole for Ewing's cruise, and almost all were supposed to be melted and cast into bombs.

"I feel sure we can carry out the operation without hazard," wrote Ewing when Henry Bigelow choked on the idea of melting TNT aboard. "However . . . , I will plan to do it [the casting] in the woods."

But as usual, Ewing and his crew arrived in Woods Hole with barely enough time to stow their incredible collection of gear aboard the ship before sailing and there was no time to make bombs in the woods. Thus in July and again in September of 1937 Ewing took *Atlantis* out across the Gulf Stream, and while the ship cut gracefully through the glittering sea, he and his students melted TNT in a steam cooker on the port side and poured it into lengths of steel pipe on the other side. A time clock went down to the sea floor with each bomb and geophone, but the system was too complicated to work well. Frequently the bombs failed to go off, and just as often the cable with all its gear dragged across the sea floor, preventing any record from being made. No useful results were obtained from this cruise, but Ewing clung tenaciously to the belief that seismic refraction lines could be run in the deep sea.

In the winter of 1940, just before World War II temporarily stopped his investigations, Ewing tried again, this time with an impressive load of free-falling bombs and instruments.

Atlantis sailed at 1:00 P.M. on January 17. The sky and sea were both a leaden, unreflective gray and the air was so cold that it was painful to work on deck. Immediately after breakfast Ewing appeared on deck with an armload of equipment. His baggy pants were rolled to his knees, his socks had slipped down into scruffy leather shoes, and since his hands were full, he used one foot to dislodge Minnie the cat, queen of the ship, from her bed atop the TNT. With his most enthusiastic graduate student, Allyn Vine, and several others, Ewing began to assemble the free-falling bombs.

Each of the devices consisted of a gasoline float, a bomb, a time clock to detonate it, bags or blocks of ballast to take the whole rig to the bottom, and a releasing mechanism made of a salt block. When the salt dissolved, the bomb and float would be separated from the ballast. At 11:00 A.M. the first of these devices was carefully dropped over the rail. As the ship moved over the choppy sea, the flag that topped the float could be seen intermittently as it sank beneath the waves. Two more bombs were deployed about a mile apart. Once the deck was cleared of explosives, the three heavy and unwieldy seismographs were unlashed and wrestled over the rail. They too were fitted out with ballast, salt blocks, and floats.

The bombs were scheduled to go off shortly after noon, and although no reverberations were felt aboard *Atlantis,* the first of the flags was spotted at 12:40. In a rising sea McMurray skillfully maneuvered the ship alongside it and within minutes it was back on board. The rest of the equipment was eventually hoisted onto the pitching deck and by eight that night the ship was hove to in a gale.

"Called at 1:00 A.M. to make a station," wrote Vine three days later when the weather finally settled down. "Lost two new buoys. Others came up."

Gales and stations alternated evenly for the next week, and at the end of this time the ship anchored in deep water on the Blake Plateau, off Florida, to make current measurements. Almost predictably the wind rose and the ship was blown off her anchor station. In disgust, Ewing agreed to run for Jacksonville.

The weather had been so thoroughly bad on the first half of the voyage that Ewing still had gear, including an experimental deep-sea camera, that he had not tested. Consequently, while the ship was in drydock having her bottom painted and warning lights installed in the heads (don't flush when the red light is on because a plankton net is over the side), he and Vine tested their equipment in the shark pool at Marineland. They wanted to get the light on the camera to come on only when pictures were actually being taken; once this was accomplished, they were satisfied that the device would work to a depth of 3,000 fathoms. In the past, undersea cameras had been used mostly to photograph fish and corals in forty or fifty feet of water, although Otis Barton (William Beebe's colleague) had photographed a spot of light at close to 100 fathoms, and a Frenchman had photographed a target hung in front of a camera at about 1,300 fathoms.

Atlantis sailed from Jacksonville on February 23, and after two days of fine weather encountered the same winds and heavy seas that had dogged her passage south. Four days out, in a near gale, "Bosun Carlton Ferns fell from the mizzenmast [where he had been mending the trysail] and landed on his back across the hydro winch," wrote the mate in the log. "Received bruises, puncture in left thigh." He had also, although no one knew it at the time, fractured a vertebra.

Ewing was again trying to sandwich his stations between violent squalls, and on March 1 he considered the weather just settled enough to try the camera. Like a bomb, it was sent down with a float, ballast, and salt release, but unlike a bomb, it was supposed to stay on the bottom for several hours. Every ten minutes a battery-powered light would come on for seventeen seconds and during this time three photos would be taken. The camera was dropped over the side shortly after breakfast and triumphantly relocated at five-thirty that afternoon. It had made a round trip of some six and a half miles, but the pictures were blank because by accident the storage battery for the light had been filled the night before with Formalin instead of distilled water.

The camera was tried twice more, and at last returned with the first photographs of the deep-sea floor. When developed they gave a cloudy view of a sandy ocean bottom with sketchy shadows that probably indicated ripple marks. Later Ewing demonstrated that such marks were common in deep water, a surprise to most geologists, who had considered the deep sea devoid of currents strong enough to produce marks of any kind.

Atlantis continued to work her way northward toward Woods Hole, and not far off Long Island the rain that had accompanied every stretch of bad weather turned to a light snow.

"Down all sail. Power only," ordered McMurray at three in the morning as the snow closed in and fell with a quiet hissing sound upon the water. The snowfall grew heavier the next day, and Mc-

Murray, unwilling to take his ship blindly into Vineyard Sound, stood far off Gay Head while the snow swirled across the decks and drifted into the lee of the deckhouses.

"We start in at 6:00 P.M.," wrote Vine as the sky finally began to lift. Early the following morning the ketch slipped past the islands of Cuttyhunk and Nashawena and headed toward the dark and wintry outlines of Pasque and Naushon.

"It looks like breakfast in Woods Hole. Hoorah!"

Ewing made three more cruises aboard *Atlantis* in the summer and fall of 1940, and on each set out his free-falling bombs (and his camera until it was lost). The four cruises, with all their complicated timing devices and free-fall instruments, had yielded only two records.

"The surprising thing is that in 1937 [and 1940] none of us realized that these heroic expedients were unnecessary," wrote Sir Edward Bullard, a geophysicist who had accompanied Ewing in 1937. "All that was needed was to put the instruments and the explosives near the surface of the sea and to treat the water as another layer in the problem." And this Ewing did with great success after the war.

Late in September of 1940 Doc and his students left the Institution and returned to Lehigh for the academic year. Within a month they were asked to come back, not to extend their seismic work but to join the Institution's first full-time staff, whose members were assembling to work on military projects.

The war was on. The Bigelow era was over. That canny New Englander had passed the directorship of the Institution over to Iselin at the beginning of the year and had returned to work full time at his office at Harvard's Museum of Comparative Zoology. Within the limits of his strength and the Institution's budget he had created the well-balanced body of fundamental knowledge he had envisioned in 1930. Using *Atlantis* as their major tool, his scientists had learned to make good physical-chemical profiles across the Gulf Stream, had formulated a general description of the water masses moving through the western half of the North Atlantic, had investigated sediments and submarine canyons, had described and quantified important communities of plankton, and, under Ewing, were developing new geophysical techniques of great promise.

"These were indeed exciting days," Iselin said. "We were skimming the cream."

6   The War Years

Whether or not any of this is a good thing to be involved in I do not
know. My feeling is that if you find yourself in a war as a scientist
you should do your best to help win it.

— Columbus Iselin

Hardly had the National Defense Research Committee been
established in the nation's capital when Columbus Iselin urged its
members to use the facilities of the Woods Hole Oceanographic Institution. In response, the committee awarded two contracts to the
Institution in the fall of 1940. One was for a study of the transmission
of sound in seawater; the other was for the development of undersea
instruments. Both were really extensions of work that the
Oceanographic had been doing informally for the navy since 1937,
and both changed the character of the Institution and the operation
of *Atlantis*.

The Institution's cooperation with the military had begun in the
spring of 1937 at the naval base at Guantanamo Bay, Cuba. Although
Henry Bigelow had not been in favor of pursuing applied research —
"deep-sea plumbing," he called it — he had allowed Iselin to try to
discover why the sonar system of the naval destroyer *Semmes* worked
only part of the time. After two weeks' work on *Atlantis*, Iselin had
found it was the temperature structure of the water that was causing
the problems. On a hot, still afternoon, for example, the top several
hundred feet of water were often arranged in distinct layers of warm
and progressively cooler water, and when the sonar was used under
these conditions, its beam of sound pulses was bent, much as light is
bent in passing through a prism. With such distortions it was extremely difficult for *Semmes* to track a submarine. The problem was
called "the afternoon effect."

From Iselin's report and other sources, it had gradually become
clear to the navy that it was essential both for ships tracking sub-

marines and for submarines evading ships to have all the information they could get on the temperature structure of the uppermost layers of the sea. By 1940 the navy wanted "sound condition charts" based on temperature surveys of the ocean. Since the time-consuming process of making hydrographic stations with water bottles and reversing thermometers could not begin to satisfy the navy's need, a new technique for making temperature surveys had to be developed. Consequently, the second contract from the National Defense Research Committee was for the refinement of a somewhat crude instrument known as the bathythermograph, or BT, which it was hoped would revolutionize the making of surveys. Developed a few years before in Woods Hole, the BT was capable of drawing a profile of the temperature of the water as it was lowered over the side of a ship. The instrument was taken in hand by Maurice Ewing and his students when they returned to Woods Hole in October of 1940, and within several months they had redesigned the BT so that it could be used reliably from a moving ship. Almost immediately *Atlantis* was sent out to make temperature surveys.

"You are to caution your officers and men against discussing the primary purpose of this cruise," wrote Iselin to McMurray. He had already asked the scientists to sign secrecy agreements.

We would be in serious trouble should it become generally known how and why the bathythermograph observations are being made. It will be best simply to say that the purpose of the cruise is to secure data on the effect of the wind and the weather on the sea surface. Needless to add, if you should be questioned by any naval officer of a foreign country you are not to disclose to him that our work is in any way associated with the U.S. Navy. In this connection you are to see to it that no reference to the Navy appears in the log, in letters . . . , or in the ship's papers.

Several temperature surveys were made of southern waters in the spring of 1941, and as the weather warmed, the work moved north. By August, *Atlantis* was ready to survey the waters off the Gulf of Maine, and with warnings from Iselin to look out for convoys, stay clear of Canadian ports, and run for home if the situation worsened, the ketch got under way.

After a slow passage to the northeast, she began sailing a zigzag course on and off the banks. Al Woodcock was the chief scientist aboard on this cruise. When work began, shortly after midnight, his first job was to instruct three young assistants on the use of the BT. As the ship steamed on through the night, he called them together near the special BT winch on the port side and showed them the trough alongside the winch where the torpedo-shaped instrument was kept. Some spares were stored in the deck lab. The BT consisted in part of a stylus that moved horizontally in response to changes in temperature and vertically in response to changes in pressure — that is, depth. Woodcock explained that as the instrument was lowered into the water and brought back up, the stylus traced wobbly profiles on a smoked-glass slide. These profiles showed the temperature structure of the water, and this information could be used to predict how well or how poorly a sonar system would operate.

Woodcock told the three something of the instrument's history, too — how it had started as an unwieldy, box-shaped package called the oceanograph or "rat trap" and had evolved through a succession of stages into its present form. He had helped test the early BTs and had been on *Atlantis* with a new model in the summer of 1939 when a sick crewman was evacuated from the ship by a Coast Guard plane. With some difficulty the seaplane had landed near the ship, and the crewman, bundled in blankets, was rowed to it in one of the whaleboats. Once he was put inside, the whaleboat returned to *Atlantis* and was hoisted back on deck. Then, with its throttle wide open, the plane roared into the wind, took a sudden jump on a steep swell, and dived into the waves. Kelley shouted to the boat crew to drop the whaleboat and, throwing themselves down the ship's side as the plane burst into flames, they rowed furiously to the sinking wreckage. They arrived in time to save five of the eight men. The crewman and two pilots were trapped in the forward compartment and drowned.

Now, two years later, Woodcock and his assistants were using the instruments that had evolved from those early tests. At 1:00 A.M. and again at 6:00, King Couper, Sterling Martin, and M. T. Johnson fitted a glass slide into the BT and lowered the instrument as quickly as possible over the side of the ship to a depth of about 400 feet. It was something of a trick, the young men discovered, to get the instrument back again without banging it against the rolling hull or letting it foul the log (a kind of ship's speedometer) that was trailed from the rail abaft the wheelhouse. When the BT was back aboard, the slide was carefully extracted. ("One slide fingerprinted — maybe by me," wrote King Couper in his diary.) The cruise number, slide number, latitude, longitude, and time all had to be scratched through the coating of smoke and skunk oil before the slide could be laquered and filed in a wooden box.

"Almost everyone stripped to the waist for suntans," wrote

In the summer of 1939 a Coast Guard seaplane crashed on takeoff after taking a sick crewman off *Atlantis*. Three men were killed. *(Courtesy Woods Hole Oceanographic Institution.)*

Couper on the afternoon of the eighth as the mainsail was finally hoisted and *Atlantis* ran off before a light breeze. With the temperature survey fully under way, a BT was let down every hour or half hour or even every fifteen minutes, depending on the ship's speed. There were no more stations to occupy, and although this meant far less handling of the sails for the crew, it also meant constant work for the scientists. They could no longer catch a few hours' sleep between stations. The BT was used day and night, hour after hour, and for the first time the scientists had to stand watches just like the crew. It was a change in routine that proved to be permanent, for more and more instruments were designed for continuous use.

On August 10 the wind picked up from the southwest. *Atlantis* heeled over and by afternoon was doing ten knots through a confusion of sparkling water and hissing whitecaps. By evening the ship was bounding over a long swell and thunderclouds had built up on the horizon.

"Woodcock, to the thunderous approval of the elements, took a bath in a thunder shower!" wrote Couper the next day.

Martin also, judging by his appearance, draped only in a towel in the lab. . . . About 5:00 P.M. wind began to really blow. As I came on watch at 8:00 P.M. the vessel rolled to both scuppers and took considerable water over the weather side. Heavy gusts or swells or both seemed to deliberately frame up on the BT operator and the boom described crazy arcs, all apparently designed to bang the BT against the ship. Fortunately the water was fairly warm, 78°–79°, and it was not too uncomfortable to go barefoot and be soaked. . . . At 11:30 P.M. further BT work was discontinued until . . . the next day.

Two days later, as he teetered on the verge of seasickness, Couper wrote:

I was relieved to hear Al [Woodcock] say that so far no scientist has succeeded in doing much continued long-time study while on *Atlantis*. There is apparently plenty of time, yet the clock watching and intermittent sleep necessary to prepare for four [hours] on and eight off seems to prevent one from getting right down to consistent work.

[Earlier] Al stood on top of the pilothouse, stopwatch in hand, while Johnson and I lowered BTs to some forty fathoms as rapidly as possible for about half an hour. We averaged three minutes per slide and had real teamwork!

For the next ten days the ketch zigzagged and the BT was monotonously dropped over the side hundreds of times. During the day the young men listened to McMurray's tales of docking ships on the treacherous Hoogli River or compared Martin's "cute moustache" to Couper's "lip eyebrow" and Johnson's "Uncle Elmer's beard." At night they outwitted the captain by stealing into the galley after midnight and making fudge out of cocoa and condensed milk. "The sailors and scientists are up most of the nights and are always eating," wrote the cook angrily when accused by the business manager of exceeding his budget.

Three weeks into the cruise Couper was suddenly wakened from an early-afternoon nap by loud shouts that seemed to come from just outside his porthole.

"Man overboard!" several voices cried in unison.

"I slipped on some bathing trunks and rushed on deck to find activity all over the place," Couper wrote.

A trysail on the mainmast was lowered and flopped hard, the engine started, and, with all the other sails finally down, we turned back. Woodcock, on the mast, had kept the man in view, and soon we could get glimpses of him kicking along half inside the circular cork preserver tossed to him. He had fallen over the port side and actually was just outside my porthole as he drifted by, yelling (wisely) at the top of his lungs. When the ship was stopped to windward, he seized a line and clambered aboard unhurt but very embarrassed.

Then lots of things seemed to go wrong. The mizzen traveler rail gate had to be repaired. One or two travelers pulled loose on the trysail, which was unbent. Then, after things quieted down, Ernest Siverson cut his right forefinger to the bone with a knife. Sterling Martin, lowering the BT as we wallowed in the trough of a [force] 6 sea, lost the hook pole used in hauling the BT aboard. A new one on a slimmer pole was immediately split by me, and we had to reinforce the weakened handle.

By early September, as *Atlantis* slogged back and forth across the Gulf Stream, the ship's company was growing restless. Couper spent more time by himself, either riding the bow or climbing halfway up one of the jolting masts, and he was not much amused when First Officer Kelley urged him to whistle for a fair wind without mentioning that McMurray could not tolerate whistling on the ship. ("It seems," wrote Couper after the ensuing run-in, "that whistling was a signal on the old sailing ships — bosun's pipes, etc., etc. — and casual whistling was punishable.")

There were frequent scuffles in the galley, too, and once the crew's messboy caught a cup of custard in the face when he told a sailor that the cockroaches were as clean as he was. "They eat the same food!" (The ship was fumigated regularly, but a new population of roaches arrived with every shipment of potatoes.)

Atlantis finally started west on September 4 and two days later was engulfed in a "pea-soup fog . . . as we came up to the Nantucket Light Ship. Again [we had] the spine-tingling experience of hearing the muted bellow of an approaching steamer [and] the suspense until . . . the hoarse wail finally came from astern, crossing our wake."

The fog drifted away that afternoon as *Atlantis* steamed down the sound and, with the temperature survey completed, she docked at Woods Hole at 7:50 P.M.

The hundreds of BT slides made on the survey were taken into the laboratory, where photographic enlargements were made of each. One set of photos was filed by cruise number and another was put into a geographic file. In the latter all the slides made within a small area of ocean (one degree of latitude by one degree of longitude) were kept together, and it was from this file that the seasonal sound condition charts and reports on the prediction of sound ranges from BT observations were prepared. These charts and reports were dis-

Val Worthington gets set to drop the bathythermograph—or "BT," as it was commonly called—over the side. The instrument registered the temperature of the water from the surface to a depth of about 450 feet. Later models could be sent to 900 feet. *(Don Fay photo. Courtesy Woods Hole Oceanographic Institution.)*

tributed to naval vessels and used in both predicting and interpreting the behavior of sonar.

In the three months that remained before the United States was precipitated into the war by the Japanese attack on Pearl Harbor, *Atlantis* was used both as a training ship for naval officers who were learning how to operate a BT and more often as a target vessel in antisubmarine experiments. Once she chased a tame submarine in simulated action that included a burst of .30-caliber machine-gun fire and the use of TNT charges, homemade bombs, and an underwater searchlight. The temperature surveys continued.

In early December, when the United States responded to the Japanese attack by declaring war on the Axis powers, the safety of *Atlantis* became a worrisome question. Since the Germans were opening a devastating U-boat offensive off Cape Cod, she could no longer work in local waters. Iselin was in favor of sending the ketch to the Caribbean, where she could continue to make BT surveys; the ship could slip safely down the coast, he thought, if she put into port each night whenever possible. There was opposition to this plan, both because it was dangerous and because it would increase the cost of the ship's insurance, but Iselin soon overcame the objections and *Atlantis* was made ready.

All during January, as sailors and marines moved into the Bureau of Fisheries and parts of the Marine Biological Laboratory at the end of Water Street, and as airplane observation posts were set up and blackout instructions issued, preparations were made for the long cruise. Several thousand BT slides were stowed in the labs, fuel and fresh water were pumped into the tanks, and the crew was drilled in the emergency procedures to follow in case of attack. First Officer Tom Kelley had already left the ship to become an officer in the Coast Guard, and the new mate was told not to enter anything in the ship's log but the time and the weather. If the newspapermen encountered in port became inquisitive, "tell them you're collecting plankton," Iselin told him.

When all was in readiness, *Atlantis* slipped quietly out of Woods Hole bound only as far as New London. She sailed with the tide on the afternoon of February 2, 1942, and as a bitter wind rose from the northwest, she pitched into the green seas outside the harbor and turned to run with the swells down Vineyard Sound. The temperature fell precipitously, spray froze on the decks, and soon the wind was screeching through the stiffened rigging. At midnight McMurray ordered the ship hove to, and it was not until morning that she could move on again and work slowly up the Thames to anchor inside the submarine nets that guarded the inner harbor.

The next day *Atlantis* made her second short move, this time to Staten Island; several days later she stood out of New York Harbor in a gentle west wind and sailed for Newport News, Virginia. Staying within the ten-fathom line, she passed along the Jersey coast, across the offing of the Delaware, and down along the sandy banks and islands off the Delmarva Peninsula. Overhead small planes and an occasional blimp watched for submarines, and on the water freighters routinely overhauled *Atlantis* as they hurried south. Like them, McMurray listened on the radio for the latest reports of submarine sightings, and when a U-boat was spotted thirty or forty miles ahead, McMurray headed straight for it "on the theory that it certainly would have moved by then."

As instructed, *Atlantis* ran without lights. Every evening two lookouts were posted in the bow, two amidships, and two in the stern.

On February 11 the ketch sailed between Cape Henry and Cape Charles into Newport News. She stayed amid the cruisers and destroyers in that busy port for four days, and at noon on February 15 passed through the submarine nets guarding the port. Slipping by the shore patrols with a wave of her special flags, she headed for Charleston, South Carolina. It was a pleasant, unusually mild day with little wind, and the ship ran down along the coast within sight of the dunes that border North Carolina's outer banks. After supper the lookouts took up their positions as usual, but reported some trouble in trying to see through a low bank of haze that was beginning to form as warm air from land moved out across the water. *Atlantis* was due to round Cape Hatteras that night. When McMurray came up from his cabin at 8:00 P.M. he stood for a while in the pilothouse, smoking his pipe and talking intermittently with the helmsman, before stepping down into the chartroom to check the ship's position. He determined the ship's next course and *Atlantis* steamed steadily on.

An hour later a sudden shout went up from the lookouts in the bow. A fully laden Standard Oil tanker emerged from the haze and was plowing straight for *Atlantis*.

"Ship on the starboard bow!"

The helmsman spun the ship's wheel, *Atlantis* lurched to port, and in a shower of sparks the two ships struck each other a glancing blow. The bow of the heavy tanker raked the ketch's starboard rail, caught the trawling boom, ripped it loose, and rolled it between the ships. One end of the boom punched a hole in *Atlantis*'s hull just abaft her main rigging.

In an after cabin Al Woodcock was thrown violently against the bulkhead. We've hit a mine, he thought. Remembering his emergency duties, he ran half-dressed from his cabin and began to dog down the watertight doors. The ship lay dead in the water, and in the unnatural silence Woodcock realized that he could hear no sound of water rushing into the vessel. Leaving his doors, he scrambled up on the dark deck. From somewhere astern he heard a deep rumble and thought he could barely make out the dim silhouette of a large ship riding low in the water.

"Are you all right?" a wary voice came crackling through a megaphone from the dark shape. "Are you American?"

"Yes, and we're all right," answered McMurray smartly, and before he could ask a question in return the rumble turned to a roar and the tanker was gone. Muttering to himself, McMurray continued his inspection of the ship. For a length of about forty feet along the starboard side the ship was badly stove in from rail to waterline. Several frames had sprung amidships, spraying rivets into the engineer's cabin, and letters from the tanker's bow had fallen onto the ketch's deck. The hole made by the trawling boom was well above the waterline, and although Backus reported a dent in the engine-room, the machinery had not been damaged. The engine was restarted and McMurray laid a new course north, back toward the shipyards at Newport News. Lookouts resumed their positions and *Atlantis* continued on through the dark, hazy night.

"Regret reporting *Atlantis* collided with *Chester O. Swain* . . . bound for Boston," wired McMurray, breaking the usual radio silence for the emergency. "*Atlantis* badly damaged, starboard side holed above waterline. Plates, frames, and bulwarks bent. Return port. Ship apparently tight."

Although the naval shipyards at Newport News were indescribably busy in February of 1942, work on *Atlantis* began almost immediately, and within a week $10,000 worth of repairs had been hastily completed and the ship floated out of drydock.

Without further accident or adventure *Atlantis* sailed down the coast to Charleston, South Carolina, and from there to Jacksonville, Miami, and Key West. Unfortunately, the U-boats were also moving south, and McMurray was told to stay out of the Caribbean and proceed instead to the Gulf of Mexico. There, during the latter part of March and most of April, Woodcock directed an unusually complete study of the daily temperature fluctuations of the topmost layers of the sea. But when the ship put into Galveston, Texas, for fuel on the last day of April, McMurray was notified that submarines were now operating in the gulf. Predictably the insurance premium for *Atlantis* rose to an astronomical 72 percent of her actual value and the Institution had no choice but to offer the ship to the navy.

McMurray did not immediately give up his command, nor did the navy ask him to, for it could see no use for an underpowered piece of torpedo bait. For the next three months plans for the ship shifted continuously. Finally in August, with the hurricane season approaching, McMurray tried to bring his command home. He wanted *Atlantis* out of Galveston before a storm caught her in that over-crowded harbor, where no one cared about her or knew how to handle her. The ketch was too slow to join a convoy on her own, but a naval tug was found whose captain agreed to tow the ship in a convoy at least as far as New Orleans.

Early on the morning of August 27 *Henry W. Card* took *Atlantis* in tow and headed out of Galveston Bay at the end of a long line of ships. They slid quietly past the Pelican Island Drydock Company, where the night crew was still at work repairing a torpedoed tanker, past military police who leaned lazily against their guardhouses in the cool of the morning, past long gray warehouses, and finally out into the bay itself. Once clear of Galveston Island, the ships got into formation, and for several hours on that clear morning McMurray and his small crew tried to accustom themselves to the unusual sensations of being pulled through the water with no sound from their engine, no wind in the sails, and no course to steer.

As the day wore on, the wind began to blow up from the southeast, and by afternoon, when the ships were passing the Sabine Shoals, *Atlantis* was throwing spray from her bow and pitching heavily on the end of her line. Scattered clouds rolled in as evening fell and the men on board gradually lost sight of the other ships and finally of *Henry W. Card*. Lookouts were posted, and while the others slept, they watched the foaming waves wash by the ship and felt her jerk and tug at her towline.

About 1:00 A.M. the men on deck noticed the trim of the ship change slightly and a moment later the vessel lost way and began drifting helplessly with the seas.

"It's the towing bridle," shouted a lookout over the racketing wind. "It's parted."

McMurray and the rest of the crew were roused immediately and came stumbling onto the dark deck. They listened, but heard only the wind and the sea. Sending the engineers below, the captain ordered the engine started. When its dependable roar shook the ship, he rang slow ahead.

"Slow ahead," responded the engineers, moving their handle of the engineroom telegraph to correspond with his, but the ship did not move.

"Slow ahead!" McMurray shouted down the voice tube.

"What?" answered a voice a moment later in what, over the years, had become a routine exchange.

"I rang slow ahead!" he shouted again into the noisy engineroom.

"What?" repeated the engineer.

"Is — any — thing — wrong?" McMurray yelled, and as he turned

118

from the tube an engineer stuck his head up through the chartroom door.

The propeller was just barely turning, Backus told the captain, probably because it was fouled by the barnacles that had grown on the ship during their four months in Galveston. Since the sails were stowed below and parts of the running rigging dismantled, the ship was helpless. McMurray ordered her anchored while he considered what to do.

Hardly was this order carried out when several men on deck picked up the faint sound of a ship's engine, and within minutes the suspense was broken as their tug's dark form came churning through the steep seas to take them in tow again. There was no chance that the ships could catch up with their convoy, and although it seemed for a while that they might join a second group, *Henry W. Card* received a hurricane warning and was ordered to tow *Atlantis* to the nearest port. The two vessels headed north that wild night and after a rough passage of a day and a half arrived in Port Arthur on August 29. Although McMurray expected to leave for New Orleans on the following day, the navy would not allow *Atlantis* to put to sea again. Since there was no room for the ketch in Port Arthur, McMurray reluctantly agreed to have his command towed along a canal to Lake Charles, in the marshy southwest corner of Louisiana. When the Coast Guard also declined to use *Atlantis* for shore patrol, it became clear that she would stay in Lake Charles for as long as the Germans controlled the western North Atlantic.

Atlantis was made fast to a newly built and nearly deserted section of wharves along the Calcasieu River. Ahead a highway bridge with a lift span crossed the quiet water; on the other side "a regular jungle," as McMurray called it, ran for as far as the eye could see. On the wharf side of the river empty warehouses rose above the ketch; behind them and all but out of sight was a small building with a telephone and showers that the sailors could use. The town of Lake Charles was two hot miles away.

Before the ship had left Galveston the younger men on board, those who chafed at the inactivity or who had been drafted, had left, and now only a small group of old-timers and local eccentrics remained. In the engineroom Harold Backus and a Norwegian engineer, Hans Cook, who had joined *Atlantis* in the mid-thirties, kept the generator and original Danish diesel in order and played cribbage every afternoon. On deck Ernest Siversen, by now half blind and half deaf, sat in the shade and sewed awnings. Because Siversen was in such poor health, McMurray had to keep an eye on his watch, and this task cut into the games of solitaire that the captain was accustomed to playing each evening. The bosun's ailments, McMurray wrote primly, were "due mainly to his age and general habits of life, aggravated by the fact that he has been quite careless in the matter of drinking beer."

McMurray was short a first officer, and for second mate he had Henry Mandly, a former whaling skipper of sixty-two whom the captain, in his acutely pessimistic mood, considered "uneducated, no navigator, and bone lazy besides." The ship needed an officer who liked to sail, he wrote a friend, a man who could go out on a small vessel for three weeks at a time "without going into a mental decline. There are men from the State of Maine who might be interested, but we are a long way from there."

For the galley crew, always the hardest positions to fill on *Atlantis*, the captain had found two Cajuns — a cook who could neither read nor write and a crosseyed nineteen-year-old messboy.

McMurray, who had already commanded *Atlantis* for more years than any other captain had (or would), was torn between staying and leaving. Although he loved to give the impression that he had taken a great step downward when he left the Isthmian Line "to sail a little piss bottle around the North Atlantic," he admitted to his few close friends that after ten years on the ketch he was much attached to her. At the same time, he knew he could expect neither a raise nor a pension from the Oceanographic, and he was understandably anxious about the future.

In November 1942 the anxiety won out over the attachment and McMurray resigned from the position of "Master of the Research Ship *Atlantis*, without limit of time," which he had received from Henry Bigelow ten years before, to accept command of a cargo steamer on the Murmansk run. Although he survived the war, he did not return to *Atlantis*.

At the Oceanographic Institution few people had time to miss *Atlantis*. In less than a year the staff had increased from 60 persons to about 300, and the budget had taken a spectacular leap from $135,000 to almost $1 million. As the Institution grew, it spilled over into a garage and temporary buildings at the southeastern end of Water Street, and at the other end of the village the navy and Marine Corps moved into the old summer laboratories and surrounded their compounds with high mesh fences. Each morning soldiers, sailors, scientists, and secretaries poured into the village on foot, on bicycles, in

Second Officer Henry Mandly at the winch. The former whaling captain stayed on *Atlantis* all during the war. *(Henry C. Stetson photo.)*

cars, and in boats. Carrying lunchboxes or briefcases and identified by badges, they streamed past gray guardhouses and disappeared into laboratories and offices to work on classified projects. Each night many men stayed on at the lab, even eating and napping there. Like Iselin, they worked "all waking hours six days a week during the whole year."

Most of the biologists at the Institution were concerned with fouling organisms and their control, but except for two cruises in 1940 on which painted plates had been towed behind *Atlantis* and checked for marine growth, the scientists worked in harbors or right along the coast. A representative from the Bureau of Ships told Iselin much later that their work had probably saved the navy ten percent of its wartime fuel costs and saved shipyards from a hopeless pileup of ships in need of scraping and painting. At the same time many of the physical oceanographers and geologists at the Institution were concentrating on the behavior of sound in the sea. The questions raised by their initial investigations had quickly led from the development of the bathythermograph (and a BT for submarines) to the mapping of sediments, which were found to affect sound (either by bouncing it back or absorbing it), and the development of SOFAR (sound fixing and ranging), a technique that involved the transmission of sound signals tremendous distances through natural sound channels in the sea. It offered the possibility of locating downed airmen and wrecked sailors.

Also in 1942 an entirely new project, the Underwater Explosives Research Laboratory, was brought to Woods Hole from Harvard. "Rapidly I found myself running a laboratory within a laboratory," wrote Iselin.

The explosives boys took over the top floor of the laboratory and various temporary buildings . . . [and] within a month we had a small explosives brewing factory on Nonamesset Island and the boys were firing underwater charges off our wharf. Within two months they were firing them in the air and beginning to break windows at Woods Hole. . . .

For me this was an annoying and difficult administrative problem. The railroad station at Woods Hole always had one or more freightcars of TNT . . . which was enough to blow up the whole town. We had to move the TNT to a magazine on Nonamesset and then spike it with chemicals to make the mixture even more destructive.

To run these explosives back and forth, as well as to test new BTs and other inventions, the Oceanographic was gradually acquiring a fleet of small boats. There were *Asterias*, purchased in 1931 as a near-shore collecting vessel; Iselin's personal launch, *Risk*, which he used to commute to and from his home on Martha's Vineyard; and *Anton Dohrn*, the launch *Mytilus*, the schooner *Reliance*, the motor sailor *Physalia*, and a variable number of small naval vessels. The largest of these ships could work along the coast from Maine to North Carolina provided they could put in to port at night, but none could take over the blue-water work that had been done by *Atlantis*.

In the spring of 1944, with war work going full tilt at the Oceanographic, the threat from German submarines diminished and the navy told Iselin that *Atlantis* could be brought back to Woods Hole. Delighted, Iselin appointed his friend Lambert Knight master of the vessel and made him responsible for finding and hiring a crew and bringing the ship north.

Knight had been a sailor on *Atlantis*'s maiden voyage and had shipped aboard her at irregular intervals all through the 1930s. He came from Martha's Vineyard, where his family had lived for generations in fine white houses overlooking the harbor at Vineyard Haven. His great-grandfather had commanded a packet ship on the Liverpool line, other relatives had been yachtsmen and ocean racers, and, not unnaturally, Knight loved to sail. He had fussed around boats since he was seven or eight years old and had rearranged each that he owned to his liking. One had carried so much rigging that she was constantly capsizing at her mooring under bare poles. After college Knight worked on the schooner *Alice S. Wentworth* and other ships, both sail and steam. When the war began he was hired by the Oceanographic to work on *Anton Dohrn*. Now, to his joy and confusion, he was master of *Atlantis*, a vessel he had always admired.

Knight's first job was to assemble a crew. When Iselin told him that the Institution could spare only four or five men, he sought advice from the captain of the racing yawl *Manxman*, whose crew had been laid off. Knight was given their names and told to look for them at a racetrack and in certain South Boston bars. Within several weeks he hired eight or nine of these men and with them boarded a train and rode south to Lake Charles.

It was early July, a hot, muggy time in the marshy lowlands, when Knight and the others arrived. They found *Atlantis* still tied up alongside the warehouses, and she was a mess. Gray paint was peeling off her hull in sheets and rust was everywhere. Her varnish was blistered, her rubber fittings were cracked and rotten, and her decking was lined with long open seams where the planking had shrunk. There were more barnacles on her bottom than rocks in the State of Maine. Before

123

her propeller would turn or her rudder move, a diver had to scrape the animals off. Only the engineroom had been carefully maintained.

Once Chief Engineer Harold Backus and Hans Cook were convinced that the ship could maneuver under power, she was eased away from the dock. Moving cautiously through the bayous and out into the Gulf of Mexico, she steamed to one of the passes leading into the Mississippi River. She waited there until a pilot was taken on, then started up the river toward the Pendleton shipyard.

Never before in his life had the pilot taken a sailing ship up the Mississippi, and in spite of the vessel's bedraggled appearance, he was wildly excited. He insisted that Knight set at least the mizzen and foresails, and this done, *Atlantis* began reaching up the river against a strong tide under both sail and power. Old Henry Mandly, the second mate, was on duty, and without word or expression he watched one of Knight's new helmsmen following orders from the pilot, who waved and shouted and fairly danced upon the charthouse. As *Atlantis* pushed her way upriver, down came a freighter tearing along with the tide, her decks heavily laden with tanks and half-tracks. When she rounded a bend and saw *Atlantis*, she gave two blasts of her horn to signal her intention to pass starboard to starboard. At the very same instant, however, the pilot instructed Mandly to blow a single blast, signaling the ketch's intention of passing port to port. Both ships veered toward the east bank of the river and the distance between them closed fast. Suddenly, with a rattle like that of a machine gun, the freighter let go an anchor. On *Atlantis*, with no orders from the terrified pilot, Mandly grabbed the wheel and spun it hard to port. The ketch came about within a few yards of the freighter. The freighter kept right on going toward the levee and, dragging her anchor, ran solidly aground.

With *Atlantis* safely on the ways in the shipyard, Knight's new crew attacked the mold and rust and began making temporary repairs. Within a week the ship was pronounced fit enough to sail, and on July 18 she left New Orleans bound for Miami and Woods Hole.

Knight was happy to be off. He had a beautiful sailing ship manned by a professional racing crew. He had no scientific stations to interfere with the sailing. He was ready.

Once out of Miami, Knight sailed *Atlantis* hard up a narrow corridor that the navy had designated as a safety zone for coastal shipping. He ordered all sail set, and with the expert handling and trimming of his crew the vessel moved smartly up the coast. Off northern Florida she was running before the wind with a fairly heavy following sea when Knight's navigator told him that the Gulf Stream was setting the ship to the east and would soon carry her out of the safety zone. She would have to be brought about on the other tack in order to make some westing. Knight hated to interrupt such a fine sail by coming about, and he asked a mate from *Manxman* whether he could possibly jibe instead.

"Sure," answered the old Swede, seeming to think nothing of jibing the ketch with all sail set in a heavy sea. "When you ready, you yust yell, 'Yibe!'"

Knight stood a few moments at the wheel, watching the sailors run across the weather-beaten deck to remove the boom guys and man the sheets. They were ready in a minute, and when Knight saw a low spot in the following seas he shouted, "Jibe ho!" and put the wheel to port. The stern came smoothly through the wind, and with hardly a rattle or a bang the sails snapped to the other side.

"That was slick!" Knight beamed at the Swede. "That was just like an America's Cup race." And with that happy feeling of confidence *Atlantis* sailed gracefully up the coast and on July 28, a day and a half ahead of schedule, swept into Woods Hole to pick up her career where she had left it two and one-half years before.

"Never has any person or thing been more welcome," said one of the onlookers long afterward, remembering the ship's tumultuous reunion with her home port. "We shouted. We waved. There were tears in our eyes."

Before *Atlantis* could embark upon her next cruise, the damage done during the ship's long stay in Lake Charles had to be repaired. A berth was found for the ship at the Bureau of Fisheries dock, down the street from the Institution, and soon a crew was boarding her every morning to refinish brightwork, holystone the decks, soogee the bulwarks, and repair or replace the rigging. A big American flag that had been painted on her side gradually gave way to patches of red lead and then to shining coats of gray.

On September 13 the weather bureau warned that a tropical storm from the Caribbean might approach the coast of New England, and on the fourteenth the storm warning was canceled and a hurricane warning issued in its stead. Remembering the devastating hurricane of 1938 and realizing that the coming storm would strike at night, when it would be impossible to watch over ships, Iselin quickly ordered the Institution's small boats into Eel Pond. Then he called on Lambert Knight. It was a hot, still afternoon, and as the two eyed the low, overcast sky with its queer purplish hue, they reluctantly agreed that with her mizzenmast half out, *Atlantis* had neither the crew nor the stamina to weather the storm at sea. She would have to ride it out at the dock.

Knight, who was to stay on the ship with Assistant Engineer Hans Cook, Ernest Siversen, and several seamen, walked back along Water Street, crossed the stone basin that formed a portion of the dock, and checked the hawsers, lines, and even the mizzen halyard, all of which had been used to make the ship fast. As an extra precaution he ordered the ship's trawl wire payed out and rove through holes in the rock jetty. Clamps were fitted on the cable, and *Atlantis* rode at the dock like a fly in a web. About suppertime, the seventy-two-foot yawl *Saluda* came hurrying across Great Harbor and, with Knight's permission, bound herself outboard of *Atlantis*. All that could be done had been done, and when the ringing of the Angelus bell came floating across Eel Pond, it seemed a signal for the citizens of Woods Hole to withdraw behind their barricaded doors.

The storm, when it came, arrived with astounding speed. All at once the wind began to blow, the tide rose, and the rain came down in torrents. The harder the wind blew, the higher the water rose, and by 11:00 P.M., when explosive gusts boomed overhead at eighty miles an hour, *Atlantis* strained at her moorings. Peering out through the wheelhouse windows, Knight could see nothing. Rain and spray blasted the ship, and tree limbs, shingles, and slate tiles from the Fisheries roof bombarded the vessel.

Out on deck old Siversen, his boys, and several crewmen from *Saluda* struggled to adjust the ship's lines as the water continued to rise. It was impossible to hear above the wild screeching of the wind. The men clung to each other in their streaming slickers and shouted in each other's ear.

Suddenly one of the seamen stuck his face up to Siversen's. "We're moving!" he shouted.

When Siversen barged into the wheelhouse a second later, Knight was already ringing up the engineroom. The ship was working loose.

As *Atlantis* swung out from the lee of the jetty, waves came dashing over her bow and water cascaded along her decks.

"Break out the anchor!" bellowed Siversen, struggling forward against the water.

"Cut *Saluda* free!" came a frantic shout from the yawl, and in a moment men with axes were chopping away the lines that ran across her mahogany rail.

Atlantis was headed into the wind now, pitching violently on her trawl wire. Then that too gave way and the vessel was free.

Atlantis, with *Caryn* to the right and behind her, rides out a hurricane. This photo was taken in 1960, not 1944, but the method of making the vessels fast to the dock was similar in both storms. *(Courtesy Woods Hole Oceanographic Institution.)*

Unable to see, Knight could do little but try to keep the ship off the rocks, whose whereabouts he would have known on the blackest night. He let the ketch sidle over toward the muddy banks of Ram Island and at 12:15 A.M. felt her ground and lean heavily to starboard. The tide had just begun to fall. As it ebbed, *Atlantis* settled farther and farther over.

The hurricane grew stronger still, and at its worst Knight could hear a high, eerie whine above the deep vibrating roar of wind and sea. *Saluda* seemed to have joined *Atlantis* on shore. When the wind dropped to only a moderate blow and the day broke, there was *Atlantis* lying in some eight feet of water at low tide and there was *Saluda*, still attached by a single line. Around the two vessels shingles, cans, and battered mooring buoys bobbed violently in the rough chop and on the beach dozens of small boats lay where the storm had flung them. *Atlantis* was too far over on her side to rise with the next tide, and by midmorning muddy harbor water was splashing over her newly varnished rail and flooding the ship. In the village, the water receding from the main streets carried shingles and garbage pails with it, and at the head of Little Harbor the falling water revealed an unholy mess of twisted railroad tracks and fallen trees.

After the streets were cleared, but long before the electric lines had been repaired, a salvage crew was brought in to refloat *Atlantis*. Twice a day, when the tide was high, they boused the ship over and at the same time tried to winch her ahead into deeper water. At first the vessel moved only six or eight feet each day, but after two weeks of this slow progress (during which time someone had to live aboard so she could not be legally considered abandoned) she broke loose and hauled ahead 300 feet in a single day. The work went on continuously then, and on October 6 *Atlantis* finally floated free on a spring tide.

The repairs made earlier that summer were begun again, and in November the ship was sent to a shipyard near New London, where she stayed until late January of 1945. Iselin did not begrudge her these months away from Woods Hole, for he and his colleagues were still busy with war projects and had neither the time nor the money to use the ketch for civilian work. In fact, unless the navy came forward with a project that involved the ship, she would sit at the Fisheries dock until the end of the war.

In the early months of 1945 many thought the war would last at least another year, and the navy, like other branches of the armed forces, continued to develop new weapons and tactics. One of its projects was to test the "squid," an antisubmarine weapon; another was to determine the precise distance between submarine and depth charge which would result in the greatest damage to an enemy sub. During the winter, therefore, the navy requested the services of both *Atlantis* and the Oceanographic's Underwater Explosives Research Laboratory to help with these projects. With naval money and motivation, the last repairs were promptly completed on the ship and she was readied for a long spring cruise to the Bahamas.

On a warm afternoon in late April *Atlantis* embarked upon her 129th cruise, after nearly three years of inactivity. Newly painted and polished, she looked very grand as she steamed out of Great Harbor accompanied by the Institution's small boats and cheered on by a crowd that shouted and waved from the dock. Herring and laughing gulls cackled overhead, and the village fire whistle tooted as the ship got sail up in the sound and stood out for New York.

After a pleasant sail and three days at a city shipyard, *Atlantis* left New York before dawn on April 27. Skirting the moored tugboats, lighters, and police boats that later in the day would be shuttling up and down the river tending the incoming convoys, the ketch swept out to sea on the morning tide. For several hours she sailed in the lee of the land, but by midmorning she began to pick up a real screecher from the northwest. Since the wind blew off the land and did not build a heavy sea, Knight ordered all sail set, and soon the ship was charging along at thirteen knots. Her lee scuppers were a flurry of foam, the rigging creaked and groaned, the rudder shook so hard that the whole ship trembled. The ketch had rarely sailed so fast.

The route assigned *Atlantis* was considered a safe one, and Iselin had instructed Knight to run with lights at night and to post only a single lookout in the bow. Off Virginia, a sleek gray dolphin, trailing a phosphorescent wake, rolled in the bow wave and for one awful moment was taken for a torpedo. But with only that momentary scare, the ketch sailed southward without incident, rounding capes Hatteras, Lookout, and Fear on the twenty-ninth and raising the St. John's Lightship off Jacksonville on the first of May. By dusk on the following day the ship sailed into the navy base at Port Everglades, just north of Miami.

"Hot and unsettled," wrote Charlie Wheeler, a young biologist who had been working on the drift of life rafts and the effects of depth charges on submarines. "What a letdown after our days at sea!"

As the last of the scientists joined the ship and the crew took a few days off to ogle at girls on the beach in their daring new two-piece bathing suits, news of Germany's surrender came over the ship's radio.

"Doesn't seem like V-E Day or any other special occasion," wrote Wheeler the next muggy afternoon. "Everyone here is doing the usual things. . . . Pulled out at 4:15, much to everyone's relief."

Atlantis sailed across the Gulf Stream, entered the Northwest Providence Channel south of Grand Bahama Island, and ran southeast among the low green cays and coral reefs that rose beautifully but dangerously from the deep-blue water. Knight moved the ship carefully through these waters, which always seemed unnaturally clear and blue to sailors from the north, and shortly before dawn on the tenth the ketch arrived at the approximate location of her anchor station, some two miles south of the western tip of New Providence Island. To test the squid, *Atlantis* and the naval vessels with which she would work needed deep, quiet waters, and the site selected was at 500 fathoms, just at the edge of the deep depression known as the Tongue of the Ocean. The ketch let out an anchor specially rigged through her bow gallows so that she could stay in position for some six weeks of uninterrupted work, and toward evening the fleet tug *Carib* tied up alongside her for the night.

The next morning the sailors on *Atlantis* rigged awnings to shade the deck, and the work began. A reinforced metal cylinder representing a submarine hull was attached to one side of a frame and a miniature depth charge was fastened to the other side. By setting off a series of these charges at different depths and at different distances from the model submarine, the Navy hoped to learn how best to position a charge so that the explosion, as well as the bubble pulse that was produced as gas from the explosion expanded and collapsed repeatedly before bubbling to the surface, would do the greatest amount of damage.

"Got one shot off this afternoon about 600 feet down . . . and the cylinder came up crushed longitudinally like an accordion," wrote Wheeler.

All the first week as the work with model submarines continued it was "hot and still and sunny and hot!" Some hydrographic stations were also made from the anchored ship, but the main work with the squid could not be started until the frigate *Asheville* arrived, and no one seemed to know when that would be.

A routine of sorts developed during those still, hot days. Work in the labs and the painting and scraping of the ship were begun after breakfast and carried on until supper, with a lazy break at noon. The men occasionally swam with a shark watch posted, and on Sundays the two whaleboats were put over the side and raced across the broad swells. Often in the evening, as the setting sun colored the clouds, Backus and a few others would fish from the deck and a concert of sorts would be given by the second mate on his concertina, a recorder group, or Wheeler on the secondhand fiddle he'd bought in Port Everglades. The nights were beautiful. A cool breeze usually sprang up, and the men, listening to music on deck, could watch the brilliant constellations move through the rigging.

"May 16. *Carib* did some shooting with her three-inch gun this afternoon [and] steamed around at full speed for a while to break up the monotony of just drifting."

Everyone was impatient for the squid work to begin. But *Asheville* was delayed again, and this time *Atlantis* ran into Nassau for three boisterous days of liberty.

Finally the frigate arrived and the squid work began. Moments before launching, each pair of these sophisticated depth charges was preset to explode at a specific depth and was put on a trajectory that would send it close to a target (presumably an enemy submarine that was constantly maneuvering). The bulbous charges were then hurled into the air ahead of the frigate, and although the squids' trajectories through the air could be observed, no one was sure what paths they followed underwater or whether the charges really went off at the preset depths. To gain this information the men had to hang a string of sensitive pressure gauges beneath the ship. These gauges registered the depth of the explosions. Since each pair of squids was set to go off at a slightly greater depth than the pair before, the records showed a descending line indicating the squids' underwater path.

To everyone's disappointment, the first trials with the squids were duds, and the system had to be taken back to Port Everglades to be fixed.

One hot, hazy, calm, humid day followed another. A shark was caught and blown up with a small explosives charge. The squid work began again with two duds. The strings of pressure gauges languidly wrapped themselves around the vessel when she swung with the current, and when that happened — whether at 3:00 P.M. or 3:00 A.M. — it became necessary to launch a small boat and unwind the whole mess.

The squids finally began to work correctly in early June. Every hour or so a pair of charges was lobbed hundreds of yards ahead of *Asheville* while overhead moving pictures of the fast-spreading shock waves were taken by a blimp that swayed and yawed above the intensely blue sea. On *Atlantis*, scientists dressed only in shorts ran barefoot in and out of the labs to watch over their sensitive electronic gear. After a good day's testing there was likely to be a celebration in the captain's quarters.

On June 13 Wheeler wrote, "*Asheville* made one turn after a firing run that could have ended in a fine collision with the tug, but both vessels backed down hard, *Asheville* blasting away on her horn."

After a few more days of testing, another weekend rolled around and the navy ships departed for Nassau. *Atlantis* was left to ride out the empty days on the Tongue of the Ocean. From where the vessel rode at anchor, the south shore of New Providence was visible as a low green mound, and in idle moments Captain Knight had taken to scanning the island with his binoculars. He had discovered four or five small buildings at the head of an inlet; his chart suggested that they might belong to the small settlement of Adelaide. He wondered what kind of place it was.

As the weekend wore on, the captain and the second mate decided to satisfy their curiosities. They lowered one of the ship's whaleboats over the side and slipped away on a light breeze. The low green island gradually grew larger, and as they approached the shore the water broke into patches of azure, green, and deep blue. Keeping to the darker, deeper stretches of water, Knight guided the whaleboat into the bay. There were the huts he had seen from *Atlantis*, and now he could see they were built of weathered coral rock. Each had a corrugated tin roof and empty holes for windows.

Following a path that led from the beach through fringing palmetto scrub, the two men soon came upon a larger group of houses. They seemed inhabited, and sure enough, within a minute or two the residents of Adelaide were crowding curiously around their unexpected visitors. A celebration began immediately. Soon an open fire was crackling in a clearing at the center of the settlement, and dozens of freshly caught crawfish, wrapped in seaweed, were steaming in the coals. Tiger brand rum and a mild alcoholic drink called *malteea* were passed around and around the circle of men and women. As the drinks took effect a rhythmic, shuffling dance began to the accompaniment of a goatskin drum. The instrument had to be heated in the fire before it was tight enough to play; then it thumped and bumped through the long afternoon.

Toward evening, Knight and the second mate, who had joined wholeheartedly in the revelry, reluctantly left Adelaide. Accompanied by half the settlement's population, they strolled down to the beach. There was the whaleboat, right where they had left it, but to their astonishment it was filled to the gunwales with pink hibiscus blossoms. Still more flowers came raining down on the two officers as, grinning and waving, they pushed their flower barge into the bay.

"Good-bye, good-bye," they called as the mate raised the sail and the captain let the whaleboat fall off the wind. Once through the channel and outside the bay, the boat caught the wind and heeling smartly, rolled briskly southward.

Atlantis seemed to grow in size as they approached her, and she gradually changed from a white dot to a ship with men along her rail.

Within hailing distance, Captain Knight changed course to fetch up in the lee of the ketch, and as he did so he gave a shout to the men at the rail. To his surprise, no one yelled back.

"What the hell?" he said, and took a closer look.

There, standing at the rail and looking down at two unshaven officers in a boat full of pink hibiscus, was the Institution's director, Columbus Iselin. He had been ferried in from Nassau unannounced late that afternoon.

"Hello, Lambert!" Iselin finally shouted, laughing in spite of himself.

A shout went up from the men along the rail.

"What a joker!" said one admiringly.

"You wouldn't catch the Old Man doing a bloody fool thing like that," Backus replied, honoring McMurray's memory as he always did.

But there were plenty on board who were ready for such adventures, for as Backus himself had written the port captain in Woods Hole as he dourly contemplated another two weeks on station, "We haven't been into port for a month and the crew are damned restless, I can tell you. You know, it makes you feel bad when you see these bloody navy ships steam off to Nassau for a weekend liberty and we are left here all alone. You should hear some of the remarks passed. Oh, boy. I daren't write them."

It was a glorious occasion, therefore, when the gauges beneath *Atlantis* wrapped themselves around the ship again and the project was ended ahead of schedule. When a tug came alongside with the intention of raising the tangled cables and bashed into the starboard after bulwarks of the ketch instead, it was decided right then and there to start for home.

"June 28. Kept trying to imagine what it would be like to be home again," wrote Wheeler as *Atlantis* sailed toward Woods Hole, "but the concept eluded me. I had never before realized how long two months can be and how much can grow hazy in that time."

The water looked almost "khaki-colored compared to the blue we have become accustomed to," he noted on July 4 as the ketch rounded Gay Head with a fair tide and came up Vineyard Sound. "We tied up about 3:45. . . . It is good to be back."

That August *Atlantis* embarked upon two short cruises, again working on naval projects. During the second of them the Japanese surrendered and World War II came to an end. On the ship the routine was scarcely affected by V-J Day, but in Woods Hole there were great shouting and celebration. The locomotive that had hauled in so many loads of TNT whistled until its steam ran out, stoked up, and whistled again.

"It's over!" people told each other again and again, and already some of the hardships seemed more a source of amazement than a cause of regret. Soon the staff at the Woods Hole laboratories picked up a second refrain — "Where do we go from here?"

"Back where you came from" seemed to be the answer as the Office of Scientific Research divested itself of power and responsibility as quickly as it could. Seeing this and similar signs, Iselin believed that public funds for oceanography would dry up and that the Institution would again be run as it had been before the war. As if to confirm his prediction, the explosives lab moved to Silver Springs, Maryland, Ewing and his associates established themselves at Columbia University in New York, and other scientists returned to their universities. Once again there was elbow room in the laboratories, and many believed that the Institution (which, like so many organizations associated with the government, had become known by its acronym, WHOI) would operate fully only in summer and that *Atlantis* would again alternate between short northern cruises and long southern ones.

But there were at least two major reasons why the Institution could not go back to its old ways. For one thing, the science itself had changed. Before the war the goal of most oceanographers had been to describe the steady state or average condition, be it of a current, a trace element in the water, a population of lobsters, or a bed of manganese nodules on the sea floor. But by 1945 the goal had changed to following the dynamic processes that occur in the oceans. Iselin now wanted to know what interplay of forces kept a particular current flowing, for example, and how the forces and the current changed from day to day and even from hour to hour. Such intensive studies could not be undertaken by a summer institution open only three months each year.

Second, as Iselin realized with something of a shock, inflation would not allow WHOI to operate as before on its fixed income.

"It has become evident that postwar costs of oceanographic fieldwork have more than doubled," he wrote. The cost of running *Atlantis* had risen from $200 per day to more than $400. "The costs of operating our own vessels have become so high that without government subsidy we could not hope to undertake offshore observations."

In other words, *Atlantis* would work on the high seas on projects financed largely by the government — with all the problems concern-

134

ing the selection and control of research that government funding involved — or she would join the launches and cutters of the coastwise fleet. There was no doubt in Iselin's mind as to which course he preferred.

7 A Period of Readjustment

Since *Atlantis* is the only vessel in this country which from the outset was designed with oceanography as her primary purpose, the navy has a special interest in her continued efficient operation.

— Columbus Iselin

In the fall of 1945, as the wartime fences and guardhouses along Water Street were taken down and the Oceanographic Institution faced the prospect of disposing of twelve tons of classified documents, Columbus Iselin knew that an anxious period of readjustment was at hand. The question uppermost in his mind was whether or not the navy would find new reasons to support oceanography. For over a year it looked as though it would not, and during this time Iselin adroitly sandwiched as much basic research as he could into the last of the navy's war-oriented cruises. Later, as the Office of Naval Research was organized and began awarding extensive contracts, it appeared that the government had reversed its policy and that oceanography (and the operation of *Atlantis*) would be supported, albeit on a somewhat erratic schedule. The question then became: Could *Atlantis* meet the greater demands of a vastly expanded science? Viewed in terms of postwar requirements, the fifteen-year-old sailing ship suddenly looked antiquated and small — and slow.

In February of 1946 the navy asked a group from the Underwater Explosives Research Laboratory to use *Atlantis* to finish up a cylinder-damage study (a project that blew up model submarine hulls with miniature depth charges). Iselin made sure that the voyage down to Guantanamo Bay would include the making of BT slides and other observations and that the trip back would include a short but excitingly new study of the Gulf Stream.

With these objectives *Atlantis* left Woods Hole at 6:00 A.M. on a cold, clear morning in early February and with her sails full and drawing bounded down the sound with a fresh breeze on her starboard quarter. Shortly after noon the newest model of BT, capable of measuring temperatures to 900 instead of 450 feet, was broken out for use on the half hour and special log sheets for the recording of weather observations were stacked in the chartroom. But only the BT was used, for as the wind picked up and the ship increased her rolling and pitching over the icy swells, more and more of the scientists got seasick and disappeared down the circular stairs.

"Apparently the whole North Atlantic was in a bad mood," wrote Gil Oakley, a former Coast Guardsman who had recently replaced Lambert Knight as master of the vessel. "Our radioman, that handmaiden of despair, was in his element, bringing reports of storms everywhere."

For five days *Atlantis* pitched and tossed uneasily, but on the sixth she picked up the northeast trades and without further difficulty was soon safely moored in Guantanamo Bay.

The cylinder-damage work began immediately, and a routine developed whereby the ship put out from the base at daybreak each morning, ran out of the bay, and hove to while the "bang boys" worked with their cylinders and charges. By early afternoon the trades had usually kicked up enough of a sea to make the handling of explosives unwise, and the work was stopped. Back into the bay came *Atlantis* and the crew was let off for the remainder of the day.

It seemed a nice enough schedule, but actually, with so much liberty, Oakley found it difficult to get the whole crew back on board. He was continually sailing shorthanded, especially since the old bosun, Ernest Siversen, who McMurray had predicted five years before would be "hard to kill," had been examined at the base hospital and again proclaimed unfit for all but the lightest duty.

"I plan to keep him as I think he is happy as a clam and sending him back would upset him," wrote Oakley, expressing the concern that the Institution typically felt for its employees in the early years. "I don't care if he does any work, but we'd better think of some provision for him on our return."

At first, the scientists, "those eight-hour prima donnas," were reluctant to shore up the sagging schedule by testing their cylinders in the evenings, when everyone else had liberty ashore, nor was the spirit of cooperation enhanced by a self-appointed "fairy godfather" of a naval captain who was trying to put the crew of *Atlantis* in sailor suits.

Taken together, these problems reduced efficiency by 50 percent, Oakley estimated, and so the situation remained until his difficulties came to a head with a bang.

It was late in February, a mild and springlike time in Cuba, and as the ketch steamed off her moorings at sunrise Oakley congratulated himself on having a full complement of men on board. As usual, he had not been sure whether the cook would make it back to the ship on time, and he had had to send a sailor out before dawn to bring him in. "Cookie," who used no seasonings except mustard, catsup, salt, and vinegar but was a good and generally reliable cook nonetheless, was fulfilling a lifelong ambition. He had always wanted to be a fighter pilot, and now, late each night when he had had plenty to drink, he

Gil Oakley became master of *Atlantis* in December 1945, and spent a large part of the following winter in southern waters helping to take underwater potographs of minia-ture submarine hulls. *(Courtesy Gil Oakley.)*

wandered over to a scrap yard on the base where parts of old navy planes were left to rust away. He climbed into one or another of the damaged cockpits, and to the sounds of his own *rmm-rmm-RMMS* and *rat-a-tat-tats* he flew through the early hours of the morning until he fell asleep. He had been sleeping right through sailing time until Oak-ley discovered his flying field and had begun sending a sailor to bring him back to the ship.

And so on this particular morning *Atlantis* got away with her full company. Once the ship was outside the bay, the large pipe frame was rigged with a heavy cylinder, a TNT charge, and an underwater camera that was to film the destructive effect of the explosion's bubble pulse on the cylinder. All morning the shots went off as planned, but after dinner the electric detonator developed a problem and the charge would not fire. The frame was raised, and one of the bang boys, a bright, impatient young man, fairly attacked the rig with his voltmeter.

"What in hell is the matter with this?" he was heard to mumble, and in testing the circuit without disconnecting the explosive, he got just enough of a charge across the detonator to set it off.

Oakley, napping in his cabin, heard a bang as if a cherry bomb had gone off on his pillow. Jumping up on deck, he found the injured man half collapsed with shock. His shirt and parts of his pants had been blown off, copper shrapnel was imbedded in his bare stomach, and, most alarming, he had lost his hearing in one ear.

"Explosion on R/V *Atlantis*," tapped out the radio operator as the first mate got the ship under way. "Man injured. Have ambulance meet us at ordnance dock."

A subsequent radiogram notified John Churchill, the Institution's first port captain, of the accident, and although the scientist recovered quickly in the base hospital, Churchill flew down from Woods Hole to encourage stricter safety measures.

By the end of April the cylinder-damage program was finished, and the ketch sailed up to Miami to pick up Fritz Fuglister and an assistant, who together were going to experiment with a new method of studying the Gulf Stream. Fuglister, the muralist who had first joined the ship for the Georges Bank cruises, had spent most of the war years collecting BT slides and filing them according to geographic location. From his collection it had become apparent that the interesting, changeable portion of the Gulf Stream was the section above Cape Hatteras where the stream turns away from the coast and spreads out in confused and contradictory patterns. Fuglister hoped to gain a clearer understanding of this region by zigzagging inside the stream as the ship drifted along with the current. This approach depended on two advances in technology made during the war — the BT and a new system of navigation called Loran A, which depended on the reception of radio signals from stations ashore.

With both these devices *Atlantis* proceeded off Hatteras. Although the ship's company was anxious to get home after such a long cruise, a few short legs were made within the stream. With the BT going almost constantly, it was immediately apparent when the ship was about to leave the stream, and she could be turned 90 degrees and sent back across the current. Although the ship was steered as though the legs were to be straight, the Loran A system was accurate enough to show that her actual tract curved in and out like the outline of a tulip petal because the course was being influenced by the varying strength of the current.

"The gadget is elegant," wrote Oakley of the Loran, and "the others are beginning to believe it now. It only remains to teach them that it won't blow up in their faces."

This experiment in the Gulf Stream — just the kind of rider Iselin liked to attach to the navy bill — was too short to yield startling results, but it did show Fuglister that he was on the right track.

After three months and a day at sea, *Atlantis* again sailed up Vineyard Sound and, as had become routine on these longer voyages, was met by one of the Institution's smaller boats. The crew of *Asterias* hailed *Atlantis* off Tarpaulin Cove near midnight that cloudy spring night, and as the ketch lay to, the launch came alongside to receive several cases of rum — a few bottles for the director and a few for each member of the crew. Prohibition had long ago been repealed, of course, but import duties had not, and the importation of a small amount of duty-free liquor had become one of the fringe benefits of working on *Atlantis*. Once the cases had been handed down, the ketch proceeded up the sound. Sailing along the intermittent path of light that flashed from Nobska Lighthouse, she moved cautiously into Great Harbor and tied up at the dock for the night.

Early the next morning two customs officials arrived from New Bedford. The ship had put into that port for her pratique, or health clearance, until 1939 or '40, but the system had been changed to the satisfaction of both parties so that the customs officials now came to Woods Hole.

Clearing customs was a relaxed affair. The officials were invited into the captain's cabin, and over drinks he told them the adventures of the voyage. Oakley was a bit worried after this particular trip, for he and Chief Backus had each brought back a Cuban parrot without knowing the rules of import. The bright-green birds had come from a shop in Guantanamo City, and since they had apparently been quite eloquent in Spanish, Oakley had reasoned that it would be easy to teach them some English. Although his bird never learned a single word, it made a fine sight as it climbed into the rigging, screaming raucously, as the ship rolled steadily along.

Luckily, the customs officials did not object to the parrots, and after a perfunctory inspection of the ship, Oakley took them into the saloon for an early lunch and several more drinks.

Jan Hahn, a scientific assistant and photographer on *Atlantis* for many years, tries to launch a BT without letting it crash against the hull of the ship. *(David M. Owen photo. Courtesy Woods Hole Oceanographic Institution.)*

Less than two weeks later Fuglister took *Atlantis* out again to track the Gulf Stream on what became known as the wedding cruise. The intention was to sail down the coast as far as Norfolk, then turn eastward into the Gulf Stream and zigzag along its northerly edge for as long as possible before returning to Woods Hole in time for Captain Oakley's wedding.

By May 26 *Atlantis* was in the stream and the 900-foot BT was being sent down about every twenty minutes. At 1:45 A.M. the BT slide, read by the lights in the upper lab, seemed to indicate that the ship was about to sail out of the current and orders were passed along to wear ship. Back across the stream she sailed with the officer on watch taking frequent fixes with the Loran and two scientists passing constantly between the lab and the BT winch making slides. At 6:05 A.M. the ship was jibed again and still again at 12:47 P.M., 5:45 P.M., and so on through the next three days and nights. Under power and sail, and with a push from the current itself, *Atlantis* ran along at better than ten knots, zigging to the northeast for some fifty miles, then zagging to the southeast for about the same distance.

By May 28 Oakley was becoming restive and he suggested to Fuglister that the ship be headed home. But because, like Fuglister, he was fascinated by the unexpectedly serpentine pattern of the stream, he allowed himself to be persuaded to make just one more leg to the southeast before running in. Five or six hours went by, and when Fuglister came back on watch he was surprised to find the ship still steaming to the southeast and the BT still registering the warm temperatures of the Gulf Stream. By this time Oakley was on hand for every lowering, expecting each slide to indicate the edge of the stream and the start of the journey home. Seventy-five miles and *Atlantis* was still in warm water, and Fuglister wondered if great loops or oxbows might sometimes form at the edge of the current. After a run of close to a hundred miles the BT finally indicated the edge of the stream and Oakley, the bridegroom, piled on all sail, called for whatever power the tired engine could deliver, and headed home.

Looking at his accurate plot of the ship's track and the closely spaced BT slides made on nearly twenty legs, Fuglister could see that his observations of the Gulf Stream differed considerably from those made in the past.

"There is a river in the ocean," Lieutenant Matthew Fontaine Maury had written in 1855, beginning his poetic description of the Gulf Stream, and for the rest of the nineteenth century and well into the twentieth, the concept had seemed apt. Even in the 1930s, when numerous profiles of the stream made by *Atlantis* suggested that the picture of a broad, weakly flowing river be changed to one of a more meandering current, the Gulf Stream was still assumed to be broader, weaker, and steadier than Fuglister's new data showed it to be.

In addition to confirming the stream's meandering nature, his data showed that the current could wobble as much as two miles to the north and south each day as it moved out from Cape Hatteras. The speed of the current was greater than expected, too, and near the inshore side a narrow band of water moved along at four to five knots. Countercurrents of up to three knots had been encountered as well, and the last abnormally long leg of the voyage had led Fuglister to suspect that loops of stream water were forming at the edges of the current and might well break away as independent eddies, which would then slowly mix with surrounding waters. All in all, the wedding cruise had produced a much more complicated picture of the Gulf Stream than had been anticipated.

Although *Atlantis* had returned from her examination of the stream on Memorial Day, in plenty of time for Oakley's wedding, the summer season was about over for the ship. She went out for a few days under an interim master, Olcott Gates, then proceeded to a shipyard in New London for the installation of a new engine.

The ship's original engine, an early diesel rated at 280 horsepower, had never been entirely satisfactory, and in spite of excellent care by the engineers, Harold Backus and Hans Cook, it had grown progressively slower and less reliable. According to Backus, the cylinders had warped, and fuel, leaking past the pistons, had begun to explode at the base of the engine. By the time Oakley took command of the ship, the engine was throwing flame and hot oil across the engineroom, "which is not a healthy combination," he wrote to Iselin, "particularly as it is often preceded by steel inspection plates. Half speed is the best we can do."

The Institution, with its fixed income already impossibly stretched in an attempt to meet postwar costs, could not afford a new engine, just as it could not afford a new mainmast or mainsail, although dryrot had been discovered above the second spreader and the mainsail clearly needed to be replaced. Luckily a 350-horsepower diesel — intended for the Russians but never delivered — was found among the navy's surplus and given to *Atlantis*. Even then, WHOI did not have the funds to install a new tail shaft, and the coupling of the larger engine to the original drive train gave years of trouble.

The navy came to the Institution's rescue at just the right moment with the engine, and this substantial gift was actually only one of several indications that the Bureau of Ships and the Office of Naval Research planned to support oceanographic studies on a regular basis. The chief expression of what Iselin called "this rather unexpected continuing activity" was the navy's decision to sponsor an oceanographic survey of the waters surrounding Bikini atoll, where atom bomb tests were to be conducted. This project involved about forty persons from the Institution, and many from other labs as well, but did not use *Atlantis*. A second project, which did involve the ship, was a survey of the waters off the southeastern United States, from the latitude of Bermuda down as far as the Bahamas, the Antilles, and Puerto Rico. The navy was considering the area for a rocket range, and the immediate objective was to find the depth of the natural sound or SOFAR channel in the water. When this channel was located, listening stations would be set up to monitor it. The navy could then fire off rockets, each equipped with a pressure charge set to go off in the channel, and, by pinpointing the explosions, determine the splash points of the shots.

Unlike such applied work as the cylinder-damage study, which allowed the Institution to sandwich only a few days of basic research into the contract, the rocket-range survey provided the money, the time, and above all the flexibility necessary for a basic study of the ocean.

Atlantis returned from New London "without all that clatter-bang racket we had with the old hookup" and was readied for the survey. For cruise 143, Nansen bottles were loaded into the deck lab, instructions were given on the operation of a new continuously recording echo sounder, and the crew and scientific staff were introduced to the ship's new master, Adrian Lane.

Lane was fond of saying that he was "the last leaf on the tree," and by that he meant he was the last captain born on Skipper Street in West Mystic, Connecticut. In the 1800s there had been so many masters of whalers and merchantmen in the community that the very street they lived on had taken the name of their profession. All the men in Lane's mother's family had been seafarers, and several traced their ancestry back to the seventeenth-century Dutch explorer Adriaen Block (whose name was given to Block Island) and his Indian wife, Morning Mist.

As a boy Lane had sailed small boats along the indented coastline of Connecticut, and after studying naval engineering at Trinity College he had joined the Coast Guard. By 1946 he was a lieutenant with a command of his own.

At the age of twenty-seven Lane had completed his service and was discharged at the Woods Hole Coast Guard station. He had walked down along Water Street and on impulse had ducked into the Oceanographic to see if he could get a job. That had been in May, and now, in September, under the attentive gazes of Iselin, Oakley (who had replaced Churchill as port captain), and most of the seafaring community, he was standing on the wheelhouse roof calmly leading the helmsman through the difficult maneuvers necessary to clear the harbor. On this, his first cruise, he had been given instructions to keep the echo sounder running whenever possible, conduct box surveys around any pinnacles or seamounts encountered, pay particular attention to navigation, train his officers and radioman in the use of Loran, and pray for the new engine.

The ship ran down Vineyard Sound on her way to Bermuda, where the survey would begin, but she had hardly cleared No Mans Land when something did go wrong with the engine and the ship was

Captain Adrian Lane commanded *Atlantis* for almost six years. *(Jan Hahn photo.)*

forced to put about. Late that night the ketch nosed carefully into Great Harbor, and Lane, again on the wheelhouse trunk, peered ahead to catch sight of the range lights on the Fisheries building.

145

When the minutes went by and the lights failed to appear, he ordered the mate into the bow with a searchlight and called for a chart to check his course.

"North ¼ west," someone had written along a penciled line leading into the harbor, a method of notation that implies a magnetic course, whereas in fact "north ¼ west" was the true course. Consequently, Lane gave the helmsman a course some 14 degrees to the west of what it should have been, and in spite of a frantic shout from the mate, *Atlantis* slipped past the unlit black can that marked the edge of the channel and ran onto a muddy shoal. Lights flashed from the ship as Lane and the mate regarded the shore, the buoy, and the black water in between, and soon answering lights appeared on the shore. Across Juniper Point the Coast Guard cutter *General Greene* could be heard warming up, and soon she came out of Little Harbor, her powerful lights blazing across the water. *Atlantis* was taken in tow and yanked off the bank so smartly that she nearly backed onto the rocks off Juniper Point. Avoiding that hazard, Lane brought his command safely to the dock.

It was an embarrassing situation for a new skipper, but Lane had learned enough about *Atlantis* to know that minor collisions with the dock and gentle groundings were fairly common. When the engine was repaired, his pride was also, and the ketch set off once again for Bermuda.

On this first leg Fuglister and his assistant, Val Worthington, who had worked at the Institution for a few years before the war, tried to remember how the hydrographic stations had been made in the 1930s, when Woodcock, Seiwell, and Bumpus had been making so many hydrographic profiles, often single-handedly. The ship's officers had been responsible for bending on the Nansen bottles in those days, but now, with so many scientists on each cruise, enough to cover all three watches, it seemed logical for them to take over the job.

After some difficulties Fuglister and Worthington got a string of new bottles over the side, and as the winch started winding them back in, Fuglister leaned out over the platform and watched for the first of the wobbly yellow images to appear in the water below.

"In sight!" he shouted, and a moment later the bottle rose from the water in a shower of fine spray.

"Hey, Val, do you remember these things leaking like this?" asked Fuglister when he realized that most of the spray was coming from the bottle itself. Worthington couldn't remember, but Rocky Miller, a new employee who had worked with the instruments before, knew they shouldn't leak, and showed the others how to grind the valves to make them tighter.

By the time the ship reached Bermuda, the Nansen bottle problem had been fairly well worked out (although not well enough to yield accurate salinity measurements on this cruise) and Fuglister and several others returned to Woods Hole, leaving Worthington to catch water for the remainder of the voyage. The rocket-range survey now got under way and the ship began transcribing a series of north-south lines that were to cover the area. A hydrographic station was made every sixty miles. It could be seen that the SOFAR channel lay some 4,500 feet below the surface.

On November 16, the survey half completed, *Atlantis* returned to Woods Hole. As she approached the mouth of the harbor, Worthington let out a shout. The old black can off Nonamesset had been replaced by a buoy with a flashing green light. "Adrian Lane Memorial Buoy," it was dubbed, and so it remains.

Geologists at the Institution resumed their prewar studies on cruise 144, a two-and-one-half-month survey of the western Gulf of Mexico. The Geological Society of America rather than the navy contributed funds in this case, and the main objectives were to determine the texture and organic content of surface and near-surface sediments to see if the sediments presently being laid down were different from those deposited during the ice ages.

Atlantis put out from Woods Hole on January 2, 1947, and in less than three weeks arrived in Galveston, Texas. There Henry Stetson and Fred Phleger joined the ship and the coring began. The two men hoped to run twelve traverses across the continental shelf — several extending into the Sigsbee Deep — and to take short or long cores every two to five miles along the entire track. Like the other investigators at WHOI, the geologists had a new instrument to try, in this case the Stetson-Hvorslev corer. Having found that the simple gravity corers used in the early 1930s could not penetrate hard sediments and that the Piggott gun corer was dangerous, Stetson and a Dane, Juul Hvorslev, had designed a free-fall gravity corer. A ten-foot brass-lined pipe was surmounted by a weight to which was attached a pilot weight that hung below the core barrel. When this smaller weight hit the sea floor, it released a large bight of slack cable and let the corer fall freely onto, and with luck into, the sea floor. The advantage of this design over the old gravity corer was that it allowed the core barrel to hit the sea floor at a much greater speed. The

A core barrel is laid carefully in its rack on deck. *(Courtesy Woods Hole Oceanographic Institution.)*

device had actually been designed just before the war but had hardly been used except for a few trials. A much shorter instrument of similar design, the Phleger corer, was also on board.

Atlantis cleared Galveston Harbor in a dense fog on the morning of January 25 and immediately began heaving to for bottom samples at two-mile intervals. The Phleger corer, run on the hydrowinch, splashed over the side, dropped through the shallow water, sank into the sediment, and filled up with several cupfuls of sandy ooze. Up it came, and as the ship got under way again, the sample was extruded and stored in a wooden rack. Stetson took a small bit of each sample into the deck lab for a cursory inspection under a microscope and noted his preliminary findings.

After supper the time came to try the new corer, and it was

lowered into gentle swells that barely rocked the ship. To Stetson's and Phleger's delight, the apparatus worked just as intended, and the corer returned to the deck full of sediment.

This success was repeated again and again as *Atlantis* continued her sawtoothed pattern of stations, and Phleger attributed much of the good work to the competence of the crew.

"The ship is in perfect condition, as far as I can tell, and is expertly handled. I have never seen it so well run . . . and I have nothing but praise for the excellent way they [Lane and the crew] are doing their job. The morale is high on the ship and in the scientific party."

Sharks were frequently caught from *Atlantis*. (George L. Clarke photo.)

In fact, the work was so successful and the weather so generally settled and the food so acceptable that monotony threatened to become a problem. At first, Lane was urged to allow afternoon swimming parties, but when he refused because of sharks, the sailors turned to fishing and were soon devising unusual ways of killing the sharks they caught. Arvid Karlson, an enormous Swede who had signed on *Atlantis* after the war, perfected a technique of lassoing them by dangling a baited line in front of a submerged noose, then jerking the noose tight as the shark stuck his head through to reach the bait. Sharks thus caught were usually shot. (Nearly everyone aboard carried a gun.) Other sharks were caught on great steel hooks, and once, as one dangled at the side of the ship, its toothy maw gaping upward, someone poured a glass of formaldehyde down its throat. The shark went berserk on the instant. It thrashed and pounded against the side of the ship, bent itself double, tore the hook out of its mouth, and floundered off astern in agony. The experiment was not repeated.

The other diversion — although only for scientists and officers — was parties, and at the height of a particularly fine one, as the ship pitched through a blow from the northeast and the men in the saloon laughed and banged on the swinging table, Phleger was inspired to send a radiogram to the Institution: "Wind 7, sea 6, sobriety 0."

The administration was not amused. The message, they pointed out to Lane, was doubtless picked up by many ham and marine operators that night, and, in conjunction with a recent newspaper article that erroneously declared *Atlantis* to be "seeking oil in ocean depths," it did not seem that the Institution's image was being well served. That WHOI was gaining a conscience and an image at all was a further result of its wartime expansion and its new position of responsibility as recipient of large sums of public money.

"The doctors practicing in Woods Hole are much concerned about the slightly crazy people we employ," wrote Iselin, who had always delighted in eccentricity and who maintained that "you have to be queer in the first place to go into science."

In contrast, the men who increasingly managed the Oceanographic hoped, as one put it, to "weed out the wild ones."

Wild or not, cruise 144 was a tremendous success. Five hundred and fifty short cores and 100 long ones (many a full ten feet) were obtained from the 551 stations occupied. In addition, a BT had been sent down at five-mile intervals, interesting observations had been made with the continuously recording echo sounder, and seismic reflection shots had been set off.

Instead of returning to Woods Hole, the ship put in at Miami, and there the cores were loaded on a truck bound for the Oceanographic and the geologists replaced by a group from the Underwater Explosives Research Laboratory, who made a one-week cruise in the immediate vicinity. Back to Miami came *Atlantis* and again the cruise number and the scientists changed. A fourth cruise, 147, took the ship back to Woods Hole under the direction of Worthington, who was still working on the rocket-range survey.

This new practice of breaking a single round trip into several distinct cruises became popular at the Institution, and, like the launching of truly multipurpose cruises and the favoring of power over sail, resulted from a desire to do more work in less time. More investigators were available to serve than there had been before the war, and the cost of buying time on *Atlantis* was now so high that a scientist wanted to get a maximum amount of data in the shortest possible time. Balancing the higher cost, which had risen from $200 to $450 per day, were all the new kinds of work that could be done from the ship. What with the new instruments and the new drive for efficient use, Iselin estimated that each day at sea in the postwar period accomplished about as much as a week's work before the war. For the time being, at least, *Atlantis* was still holding her own.

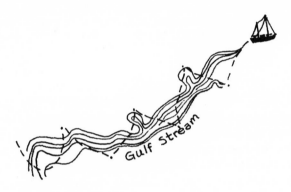

8 Maurice Ewing and the Mid-Atlantic Ridge

He had a profound effect on the success of this laboratory. He arrived here first as a very young professor. . . . He brought with him several Lehigh students, and the place has never been the same since. They literally worked night and day and seven days a week.
— Columbus Iselin

In the spring of 1947, while the rocket-range survey was being completed and Fritz Fuglister was finding one of the independent eddies in the Gulf Stream that he had predicted, Maurice Ewing resumed the seismic work he had begun before the war. He wanted, he told his graduate students at Columbia, to use the new instruments and techniques developed during the war to examine a large offshore area, the more spectacular the better. The plan was a perfect expression of Ewing's formula for scientific success: go somewhere where no one has gone before with an instrument no one has used.

The unexplored territory that Ewing chose was the Mid-Atlantic Ridge, a chain of submerged mountains whose outlines had been roughly ascertained over the past seventy-five years but whose actual length, breadth, height, composition, and formation were virtually unknown. Ewing attacked this mammoth project with characteristic energy, and the first of what became a series of cruises to the Mid-Atlantic Ridge got under way on July 16, 1947.

Early that morning the last of the hydrophones, camera equipment, and core pipe were stowed on *Atlantis*, and a navy truck rumbled out onto the dock with a ton of TNT. As blocks of explosives were passed onto the ship like buckets in a fire brigade, lines were made fast around twenty drums of fuel oil loaded on deck for the long voyage. With everything in place, there was hardly room for families and friends to pick their way around the deck in the general tour of inspection they always made before an important departure. Columbus Iselin was particularly excited about this sailing.

"This may be a long and tiring voyage," he told Captain Lane, but "the scientific importance of the program is greater than any other you have yet undertaken."

Ewing, too, was urging the ship's company on both by word and by action. "We're paying for this ship twenty-four hours a day," he said more than once as he packed TNT into a wooden magazine on the port side of the deck lab, "so damn it, we're going to *work* twenty-four hours a day!"

The tension and excitement that Doc inevitably transmitted to his colleagues spread from them to the officers and crew of *Atlantis*, and when the implications of all the high-pressure preparations finally filtered down to the ship's galley, the cook laid down his pots and pans and quit. Work on a two-month cruise with all these crazy people running around twenty-four hours a day with bombs and depth charges? Not him.

There was less than an hour till sailing time when a distraught Captain Lane took the problem to Iselin. It was handed right back. With the ship leaving and the cook staying, Lane felt as if he had one foot on the dock and one on the rail. Although he was badly needed on deck to oversee last-minute preparations, he hurried below and persuaded the slow and rather pasty-faced Cookie to leave his gear on his bunk and accompany him across the street to the Cap'n Kidd bar. There, over as much rum as the captain could get down the man in an hour's time, Lane argued, pleaded, threatened, and cajoled until the cook was so befuddled and worn down that he allowed himself to be led back across the street to the ketch.

With the cook safely aboard again, Lane strode in and out of the wheelhouse, trying to check a dozen items with his officers all at once, and out of the corner of his eye he could see Doc restlessly waiting for the ship to sail. In an unusually tight voice for a man of great outward calm, the captain fired his final orders at the crew. The last of the visitors were fairly pushed ashore, the gangplank was lifted away from the ship, and the docking lines came snaking in through the hawser ports. Lane leaped to the wheelhouse trunk and with uncommon dispatch directed *Atlantis* out of Woods Hole and away on her first cruise to the Mid-Atlantic Ridge.

Once out of Great Harbor, *Atlantis* motored sedately down Vineyard Sound. A gentle breeze blew against her from the southwest, and as the sun fell below a layer of scattered clouds and prepared to

Atlantis steams out of Great Harbor past one of the old island ferries on her first Mid-Atlantic Ridge trip. *(Claude Ronne photo. Courtesy Woods Hole Oceanographic Institution.)*

set over the Elizabeth Islands, a late and inauspiciously mediocre supper was served in the mess.

The first and least exciting leg of the journey took the ship to Bermuda (for rum, it was often said), and in spite of Ewing's injunction "Never do nothing with ship time," only a few plankton tows and hydrographic stations were made. Most of the time was spent fussing over incompletely built instruments and repairing "the scientists' powder room."

Ewing, who had not used *Atlantis* since 1940, found the ship surprisingly unchanged. She had a more powerful engine, a Loran navigation system, a more sophisticated echo sounder, and a half-dozen minor improvements, but all the rest was the same. The work to be done from her decks was not.

On July 24 the ketch left the steaming summer heat of Bermuda and the real work of the cruise began. A small seamount rising to within a mile of the surface from the sea floor east of Bermuda was picked up on the echo sounder, and the course was altered to make a quick survey of the area.

While still over the seamount, Ewing decided to break out a Stetson gravity corer — "the cookie cutter," he called it — and give it a try. He had never cored before, but Stetson's assistant, David Ericson, was on board to show him how the barrel was fitted onto the

weight, hoisted over the side, and lowered away. A small crowd
gathered around the dynamometer, and while the cable ran snapping
and crackling off the big winch drum they waited for the flicker of
the dynamometer's needle to tell them the corer had hit bottom.

"There!" shouted several voices at once, and the command was
given to reel the barrel in. Several men left for a quick supper, but
most were back at the rail when the corer came on board. The core in
its brass liner was extruded from the barrel and Ewing could see that
it consisted of a coarse, cream-colored ooze on top of a fine-grained
white sediment. The core, Ewing hoped, contained at least the last
several chapters of the earth's history — a layer of recent ooze
underlain by the fossilized remains of progressively older animals
whose identification would reveal the depth, temperature, and other
characteristics of the ancient sea in which they had lived.

"Well, some people, you know, never get over their first anything,"
said Ewing later, remembering those promising layers, "and as it
happened, that core was one of the best of my life."

When Ericson studied the core, however, he found that the pro-
gression from recent to ancient sediments was badly interrupted.
The top eight inches were recent, but all the rest had been laid down
during the Eocene, some 40 to 60 million years ago. In what was
imagined to be the utter calm of the deep sea, what could have

happened to the sediments of the intervening eons? Ewing's response was to take more cores; but not immediately, for the seismic program, the most important part of the cruise, was ready to begin.

At first, as *Atlantis* ran southeast toward the Mid-Atlantic Ridge, each of the seismic reflection shots required that the ship be hove to. A hydrophone was suspended beneath her stern, a charge rigged and exploded, and finally a record made. The whole procedure took about an hour, and when it was combined, as it often was, with the even more time-consuming hydrographic stations, there was time enough in a twenty-four-hour period to get off only three or four shots. (That the shots were going off at all in the deep sea was due to the realization — not uniquely Ewing's — that the water itself could be treated as the topmost layer in a "cake" whose bottom layers consisted of sediments and rocks. The explosives and recording instruments didn't have to lie on the sea floor, as had been thought before the war, but could be used in a much simpler manner at the surface.)

Shortly after midnight on July 28 the ketch sailed out over what eventually became known as the abyssal plain, and the image of the sea floor made by the echo sounder changed from one of irregular hills and ridges to one of uncommon flatness. The depth was 2,900 fathoms, about three and a third miles, and as minutes passed, then hours, this depth remained remarkably constant. There were occasional bumps and hollows, but Ewing knew that deviations of less than thirty or forty fathoms could well be due to fluctuations of the ship's electric current, as when the refrigerator was opened and its motor came on.

Some two and a half days later, as the weather grew increasingly hot, the stylus on the echo sounder ceased sweeping along at 2,900 fathoms and rose to reflect the presence of low, isolated hills. Higher, steeper rises soon followed, divided at intervals by wide terraces. Together these configurations made up the foothills of the Mid-Atlantic Ridge. As the echo sounder in the stuffy lower lab drew its exaggerated profile of jagged pinnacles and steep gorges, Ewing became increasingly excited.

"Six-pound boy 1 Aug. Both dandy," read the radiogram announcing the birth of the third Ewing child.

"Operating on Ridge. Morale high," he sent back. He added that the terraces he had discovered "suggest great series of extremely ancient beaches. Will confirm or refute by reconnaissance. . . . Beach theory revolutionary if confirmed but confidential now."

Ewing was in full cry. Every rise and fall of the echo sounder, every seismic shot, every core and dredge haul promised to give new information, and surely one of these measurements would lead to a big discovery.

As Ewing worked with every sweating watch, rarely sleeping more than an hour at a stretch, *Atlantis* ceased her journey southeastward and turned north to cross and recross the mountainous crest of the ridge. When the decks had been cleared of oil drums, the big rock dredge with its chain bag was broken out and was used alternately with the corers.

On August 7 the echo sounder outlined a spectacular peak that rose from the floor of a deep gorge to a height of nearly two miles. Ewing wanted a piece of that rock, and although an easterly wind blew fitfully over the gray water, slapping waves against the ship, he called for two corers and sent them down simultaneously. It was a long round trip, and the corers did not return until late that night. To Ewing's intense disappointment, neither had hit bottom; one was immediately sent down again. This one hit. Ewing saw the strain on the dynamometer jump as the corer pulled out and the long return trip began.

Standing on deck near the winch controls, Ewing suddenly saw one strand of the cable break and begin to unravel. Although breaking lines and cables had killed sailors on oceanographic voyages since *Challenger* put to sea in 1872, Ewing characteristically figured that "the place it breaks is the safest place to be." So thinking, he took a roll of tape out of his pocket, walked over to the straining cable, and taped it up.

An hour or more later, as a rim of gray light spread along the horizon, the corer came back on deck. Its tube was badly bent, evidence that it had smashed against a hard, rocky surface. Caught in its mangled end was a small piece of freshly broken igneous rock.

This was what Ewing had hoped to find. The ridge was not buried under sediments. It was accessible, and it was volcanic.

By breakfasttime the rock dredge was on its way down to the two-mile mountain, and by suppertime more than 500 pounds of volcanic rock had been brought to the surface. Ewing gave those lumpy rocks a brief but loving examination, and before a late-night lunch was set out in the crew's mess he was working on an altogether different project.

As *Atlantis* left the mountain and continued along the ridge, the objective became to fire off seismic shots in rapid succession to gain an idea of the amount of sediment, if any, covering the mountainous terrain below, and to look at its stratification. Reflection shots made from a single ship might provide this kind of detail if they could be set off rapidly enough.

Ewing and his student, Frank Press, set to work on a method that would eliminate the need to stop the ship for each shot. They discussed the problems on deck in the hot afternoons, argued briefly in the lab over warm beer (the refrigerator wasn't cooling well), and went over their plans again in the saloon as they poured A-1 sauce over half-spoiled meat that Ewing maintained would glow in the dark. Soon there were trials, and errors too. One night a charge was inadvertently fired a few feet below the ship's rail and the tremendous explosion, followed by the sound of shattering glass, brought the entire ship's company up from below. No one was hurt, and the ship's hull was only slightly dented.

Finally a new method was worked out, and on August 14 some sixteen shots were made between 3:00 A.M. and 9:30 P.M. A three-man seismic crew handled the entire routine. As *Atlantis* rolled rhythmically along through the warm, starlit night, the man in charge of explosives moved quietly past several motionless figures who had left their hot cabins to sleep outside on shrouded piles of TNT. Opening the magazine, he took out a half-pound charge. After tying the TNT to an empty beer can, whose single hole had been neatly soldered shut, he walked over to the cap locker, pulled out a fuse and a blasting cap, and inserted them in the charge. Then he carefully lit a cigarette and positioned himself on the fantail not far above the luminescent wake that spread out behind the ship. Beside him was another man who, with a great bight of the hydrophone cable in his arms, was ready to throw the slack over the stern. (This procedure would allow the trailing hydrophone to rest quietly in the water for the several seconds during which the record was actually being made.) A third man, the recorder, waited in the deck lab, one arm stuffed into a box ready to crank through a reel of paper and his head stuck under a hood as he waited to watch the galvanometer trace.

"Ready whenever you are," called the explosives man to the officer on watch, who in turn shouted down the voice tube to tell the engineer to idle the engine.

"Engine stopped," came the reply. As the ship glided on through the water, the end of a cigarette glowed brightly for a moment and the fuse began to sputter.

"Over!"

The glinting beer can with its sparkling fuse described a long arc and splashed into the phosphorescent wake of the ship. Thirty seconds later it exploded with a sonorous boom.

In the deck lab the recorder had been watching his stopwatch, and just before the charge exploded, he turned on the movie camera that would film the wiggle trace and yelled, "Slack!"

The excess hydrophone cable was heaved over the side, and as the ship, moving more slowly now, ran out the slack, the stationary hydrophone picked up the sound from the explosion as it ricocheted off the sea floor as well as the sounds that passed through the sediments to some deeper horizon and came bouncing back a few seconds later.

"O.K.," called the recorder, his voice still muffled under the hood, and *Atlantis* was on her way again after less than a minute's delay.

As the men on the fantail bent over the rail to pull in several hundred feet of hydrophone cable with an occasional jellyfish wrapped slickly around it, the recorder took his fourteen or fifteen seconds' worth of film out of the camera and went below to develop it. By the time this task was accomplished, and the men camping out in hammocks and the empty lifeboat were soundly asleep once more, it was almost time to repeat the procedure. The team made four shots in as many hours, then turned the job over to the second team. After a meal and a nap, the first team came back on deck to make another round of shots. It was hard, monotonous work, and years later, with the advantage of hindsight, even Ewing could see that it hadn't been worth much. Layers or "horizons" were indeed apparent, but the shots were so widely spaced that it was impossible to know which layers shown by shot A corresponded with those revealed by shot B, and so on. Without such interpolations, no picture of the arrangement of sediments could be drawn. (This realization soon motivated the design of instruments that used electrical or mechanical noisemakers instead of explosives. With them powerful pulses of sound were more safely and regularly generated every few seconds.)

In spite of the stepped-up explosives work, there were still moments of relative calm on this almost frantically active cruise. When the dredge was down and the shooting had stopped, sunburned men badgered Lane to let them swim. Only one man was allowed in the water at a time, and even this cautious routine was halted when a sneaker dropped by a swimmer climbing back on board was grabbed by a hammerhead shark that had been hiding under the ketch.

"The dredge made a great fuss coming up, bumping along the bottom and swinging the ship completely around several times,"

A seismic shot goes off astern. The photo was taken on a later cruise. *(Courtesy Woods Hole Oceanographic Institution.)*

Maurice Ewing looks at the seismograms or "wiggle traces" that will help him to estimate the arrangement of sediments below the sea floor. *(Courtesy Woods Hole Oceanographic Institution.)*

161

wrote one of the men in his diary on one of the calmer days. "The strain on the wire went up by jerks as great as 1,000 pounds."

That haul, made on one of the beachlike terraces (which scientists today think have nothing to do with beaches or any other continental configuration) produced several hundred pounds of manganese-coated rocks. Later, one of the rare hauls made by the Blake trawl yielded 200 species of animals, a lump of coal, and two clinkers.

Atlantis worked east for a few more days, sailing away from one gorgeous red sunset after another, and on August 18 put into Ponta Delgada in the Azores. More than half the work of the cruise had been completed. After inadvertently refueling with bunker oil, which covered every inch of the ship with greasy black soot, the ketch recrossed the Atlantic with only a few detours to study further portions of the ridge. Seismic reflection shots and coring continued at a dead run, as they had on the eastward crossing, and it was not until noon on September 13, when the ship sailed into Vineyard Sound, that the hydrophone was finally pulled aboard and the explosives put away.

"Down all sail" at 4:00 P.M., "pass Gay Head Light" at 4:39, and "all fast" by 6:05. As Iselin had predicted, it had been a long and tiring voyage, but as he had also foreseen, it had brought back immensely exciting information that added new detail to a previously bland and generalized picture of the ridge and deep ocean basins. It also raised at least a dozen puzzling questions.

Almost exactly a year later, Ewing, his brother John, and several students borrowed *Atlantis* to make a further survey of the more northerly sections of the Mid-Atlantic Ridge. Essentially the same kinds of information were brought back, but in smaller amounts owing to winch trouble, poor weather, and Ewing's need to leave the ship to be with his sick father. In consequence, a third and much longer voyage was scheduled for October 1948. Ewing, who had to remain in New York to teach, put one of his graduate students, Bruce Heezen, in charge.

Cruise 153 was a bad-luck cruise. The ship departed on October 11 and immediately ran into a week of squalls. At times the short, irregular seas became so steep that *Atlantis* could not keep sufficient way once her engine was idled for seismic shooting, and she would begin drifting backward toward the lighted charge.

As soon as the weather cleared, an airborne magnetometer was towed astern — a profile of the earth's magnetic anomalies had never

Joe Worzel, left, and Stan Bergstrom roll a depth charge over the rail. *(Courtesy Woods Hole Oceanographic Institution.)*

been made across an ocean basin — but the electrical cable connecting the instrument to the ship failed repeatedly. When the instrument was fixed, temporarily, the weather closed in.

After two weeks at sea *Atlantis* hove to in a steady rain on a gray and lumpy sea to take a deep core with an unusually long (forty-foot) corer and to send down a camera at the same time. In oilskins and heavy seaboots the men gathered on the starboard deck to rig the two instruments. They had worked themselves into a steaming sweat by the time word came to lower away. Both instruments splashed into the water and disappeared.

Although he disliked getting wet, Chief Engineer Backus came up on the glistening deck that afternoon to fret over a resistor on the heavy trawl winch that was heating up and threatening to put the winch and the coring project out of commission.

Two hours later the camera came up, and in the rain, with the ship making deep, ponderous rolls in the trough of the sea, the sailor at the winch controls did not hear the lookout shout, "Hold it!" The instrument rose out of the water at full speed and smashed into the blocks. The cable parted, and as the camera splashed back into the sea, the broken end of the cable snapped inboard and lashed the winch operator painfully across the shoulders. The camera work was temporarily halted, and within a few days Backus declared the big winch unsafe and the coring was stopped too.

Badly in need of several key parts, *Atlantis* sailed immediately for São Vicente in the Cape Verdes, but when she arrived in Porto Grande on November 3, Captain Lane discovered that the parts had been shipped to Dakar in Senegal instead. Again filling up with bunker oil, the only fuel available, *Atlantis* steamed straight for Dakar. Within twenty-four hours the parts were on board and the ketch was away. Now the seismic shots began to go off more than twice as fast as on previous cruises, at a rate of almost three an hour. Coring and camera work were resumed as well.

Four days out of Dakar a corer and two and one-half miles of steel cable were lost. This was only the first indication that the voyage back across the Atlantic was going to be as thoroughly jinxed as the eastward passage. Soon a trawl was lost, and after a rare stretch of fine sailing weather, the old mainsail blew out along its foot. The sail was so obviously rotten, and the job of sewing it up was so tedious and time-consuming, that it was stuffed away in the sail locker without repair.

On November 22, two weeks out of Dakar, an ordinary seaman complained of severe abdominal pains. Lane ordered full power and

The trawl comes up with its net torn off the head rope. The National Geographic Society helped sponsor several ridge cruises, and while this sponsorship produced great photographs and good science, it also yielded some flowery prose that embarrassed Ewing for years. (© *National Geographic Society*.)

set a course directly for Barbados, three days away, for he guessed that the man, having gone on a spree in Dakar, was in for "the same old ulcer business that he's had before."

Lane gave the seaman morphine at midnight when he was called to the fo'c'sle by the sick man's companions, who could hear him moving restlessly in his narrow bunk. *Atlantis* plowed on with only momentary stops for seismic shots until late at night on the twenty-fourth, when Chief Backus, on watch in the engineroom, felt a shock and heard the engine start to race. He shut it down.

"Something's gone wrong with the propeller," he shouted up the voice tube, and in less than fifteen minutes the jumbo was set and the sailors were hoisting the jib and the mizzen.

"I heard a rumbling noise," Backus later wrote for the Lloyd's surveyor, "and [I] could feel the propeller turning around on the draw bar. . . . I got the captain down to listen to it, and we came to the conclusion that the tail shaft had broken."

Near midnight the captain gave the sick man another shot of morphine. It was obvious that he was getting sicker, and as *Atlantis* slipped along in a gentle breeze under the cloudy night sky, Lane became worried.

At 5:40 A.M. he was aroused by a loud bang aft. Down in the engineroom the assistant engineer went immediately to the tail shaft, put his hand on it, and found the vibration gone. The propeller had fallen off.

Lane now called a halt to the seismic work and ordered the mainsail brought up from below. With leather palms and curved needles two men began to mend it stitch by stitch. At 1:20 P.M. Lane doubled the crewman's dose of morphine, but now the wind died and *Atlantis* lay becalmed. There was no engine, no wind, no mainsail, and no relief for the delirious seaman. At the captain's command, Oakes Spalding opened his radio transmitter and sent out a call for assistance. Was there any ship near that patch of flat, faintly ruffled water 300 miles east of Barbados that could take a sick man off *Atlantis*? Spalding sat in the tiny radio shack off the saloon and waited for a reply. It came from a British tanker, and yes, although they were five or six hours away, they would take the man aboard and rush him to the hospital in Barbados, where a doctor was already waiting. *Atlantis* sat all that evening on a placid sea, waiting. At 10:30 P.M. the captain administered another dose of morphine, and as the seaman sank into a troubled, mumbling sleep, the wind came up with a rush out of the southeast. What sail there was was quickly hoisted, and by the time the freighter's lights came rising and falling over the dark sea, it had become too rough to make the transfer.

Atlantis was sailing smartly now (the loss of her propeller enhanced her sailing qualities), and she spanked along until three in the morning, when in a sudden gust her mizzen ripped. Deck lights went on, the sail came down, and for the rest of the night the mate and the bosun dropped their work on the nearly completed main and began stitching the mizzen. Late the next afternoon the seam was finally sewn shut and the mizzen went rattling up the mast, followed an hour later by the main. With uncommon luck both sails and wind held through the following day, and on the afternoon of the twenty-sixth the ship rounded South Point, picked up a pilot, and maneuvered around the obstacles in Bridgetown Harbor on her mizzen alone. By evening the seaman was in the hospital. He recovered, several of his messmates suggested, with remarkable speed.

"This has been one bitch of a trip, believe me," wrote Lane to Oakley, unaware that he would have to take the powerless *Atlantis* with her rotten sails slamming through a ferocious gale on his way north in late December. "If our luck doesn't improve on the way home, there is likely to be one grand exodus. . . . The morale, frankly, is low."

Cruise 153, as exhausting for Lane and his crew as it was unsatisfactory for Ewing's group, finally came to an end three days before Christmas. While *Atlantis* went into drydock for extensive repairs, Ewing and his students moved themselves and their data into Columbia University's new Lamont Geological Observatory, a converted estate in Palisades, New York.

Since the end of the war it had seemed to many men at Woods Hole that Doc and his boys had dominated both *Atlantis* and the field of marine geology. As far as geology was concerned, this was close to the truth; but as for *Atlantis,* one of her grandest adventures occurred between two of Ewing's expeditions to the Mid-Atlantic Ridge. From December 1947 until June of the following year the ketch made her first visit to European waters since her launching, seventeen years before. She cruised the Aegean and eastern Mediterranean seas, and the voyage is still referred to with immodest pride as "the Med trip — the Med trip — the Med trip!"

9 The Med Trip

> Every mast and timber seemed to have a pulse in it that was beating
> with life and joy, and I felt a wild exulting in my own heart, and felt
> as if I would be glad to bound along so round the world.
> — Herman Melville, *Redburn*

With the sliding gangplank connecting *Atlantis* to the wharf
that cloudy winter morning in mid-December, the ship was temporarily a part of the land. She was like a small island at the end of a
causeway, and landsmen and sailors passed freely back and forth,
participating in each other's world. The usual business of topping off
on fuel, water, and stores was going on, and children, some carrying
Christmas presents they had been given two weeks early, ran among
the dwindling piles of paraphernalia still remaining on the dock.

By 9:30 A.M. sailing time was at hand, and without apparent sign or
signal, the people around *Atlantis* separated themselves into those
who were going and those who were not. The gangplank remained in
place, the men at the ship's rail were less than ten feet from those on
the pier, yet the distinction between the two groups was absolute.
Startled for a moment by the abruptness of the separation, the
groups stared at each other. Then the ship's whistle blew and the
shouting and waving began. Sailors jumped to raise their end of the
gangplank, and twice the necessary number of hands reached to lift
the docking lines from the bollards. One by one the hawsers slapped
into the water, and as they were drawn through the hawser ports, a
narrow triangle of water opened between the ketch and the wharf.
The ship was off.

Cruise 151, to the eastern Mediterranean, was to be the longest
voyage that *Atlantis* had ever made. The ship was under contract to the
U.S. Naval Hydrographic Office, and the primary objective of the
cruise was the preparation of bathymetric charts in and around the
Aegean Sea. These charts were being made so that the navy could, if

it wished, send submarines into the area to counteract Russian influence — a serious concern in 1947. For political reasons a naval survey of the eastern Mediterranean was out of the question, so the Hydrographic Office had requested WHOI to do the job while seeming to be solely engaged in a hydrographic survey.

The secondary objectives of the cruise were more purely scientific. Since the vessel would be spending nearly half a year in a little-known body of water, oceanographically speaking, it was planned that cruise 151 should be a kind of omnium gatherum, a sampling of all kinds of data. Physical and chemical sampling would be done throughout the voyage, and at intervals plankton hauls, seismic reflection shots, camera stations, wave measurements, weather notations, and many other observations would be made. It sounded like a frantic schedule, but actually it was not. On the Med trip no one was out to make his reputation, everyone liked to sail, and most considered *Atlantis* ideal in every respect: it was a relaxed expedition.

Cruise 151 required several pieces of new equipment. For the charting, which would require particularly accurate navigation in a region where Loran stations did not exist, a gyro compass and a radar set had been installed. The first would allow more accurate fixes to be made and courses laid, and the radar would be used at night, when the ship's position could be ascertained by bouncing signals off the islands that would almost always be nearby. For the scientific programs the major piece of new equipment was Ewing's newest deep-sea camera, which had recently returned from the first Mid-Atlantic Ridge cruise with only three or four creditable photos.

So equipped and so charged, *Atlantis* ran down the cold, slate-colored sound and proceeded out across the continental shelf on her way to Bermuda. Dean Bumpus, senior scientist on the cruise, called for a station every six or eight hours. Water samples were collected, different kinds of plankton hauls were made, and the camera was rigged and sent to the bottom.

The camera, which like earlier models was mounted on a pole with a single flashbulb, took only one picture for each trip it made to the bottom, and the first two had shown nothing but an indecipherable blur. The third was taken on the first night out as a fine drizzle enveloped the ship. With feelings of both anticipation and dread, Dave Owen, the man in charge of photography, took the film down to the darkroom in the lower lab. The picture had been taken in more than a mile of water. When it was developed, the single frame showed a patch of sea floor about six feet in diameter, dotted with

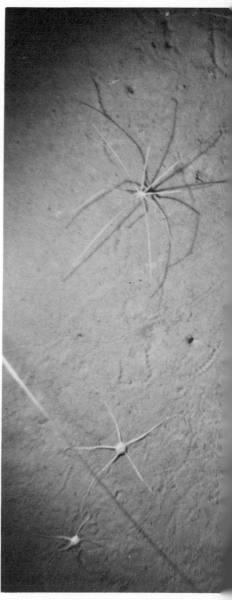

Using a Ewing deep-sea camera, Dave Owen took one of the earliest and most dramatic photographs of deep-sea life. The brittle stars and the much larger sea spider, more than two feet across, are living more than a mile beneath the surface. *(Courtesy Woods Hole Oceanographic Institution.)*

small holes and, more spectacularly, crossed by several brittle stars and a single sea spider more than two feet in diameter. This was not the first photograph of really deep-sea life, but it was one of the most dramatic, and just the kind of testing and preparation that Bumpus liked to have done before the main part of the cruise began.

On December 14 *Atlantis* arrived in Bermuda. After letting off three scientists and taking on two others, she sailed again under clear skies. The ship's company was now set for the remainder of the voyage. Lane, of course, was captain; his mates were Arvid Karlson and Donald Fay. Gus Lindqvist, a Dane with clacking teeth, had become bosun when Siversen was finally retired to Snug Harbor, a home for old sailors. Engineers Backus and Cook, having stood six hours on and six off all during the hectic Mid-Atlantic Ridge cruise, had prevailed upon Iselin to hire a third engineer before the ship went out again. A twenty-year-old from Boston had signed on.

Among the scientists, two had served aboard the naval surveying vessel *Bowditch:* Martin Pollak, chief (as opposed to senior) scientist in charge of the charting project, and his navigator, Nat Corwin. Val Worthington, Eugene Krance, Bob Abel, Frank Mather, and Dave Owen were aboard to help with both the charting and the hydrographic survey, and Dick Campbell, the Oceanographic's first electronics specialist, had been hired at a salary exceeding that of anyone else on the ship to keep the radar, gyro compass, and continuously recording echo sounder in operation. Of the entire scientific staff, only Bumpus was married, and it was said with some truth that the Oceanographic paid so little that the others couldn't afford it.

By Christmastime *Atlantis* was sailing eastward along the 31st parallel, and a leisurely routine had evolved. One or two stations were made each day, and since their positions were not critical, they tended to occur at 8:00 A.M., after the senior scientist had had his second cup of coffee. Once a hydrographic cast had been made or a photograph taken, full sail was set once more, and bowing to what were unusually steady winds for the horse latitudes, the ketch put her lee rail under and drove for Gibraltar.

For the crew of the ship the passage east felt less like the beginning of an adventure. The bosun, Gus Lindqvist, was no idler, and he had the seamen chipping and painting down one side of the vessel and up the other — on deck in good weather, below in bad. Karlson, the big Swede, also believed in working his watch.

"Loofers," he would grunt in disgust when he came upon a man with nothing in his hands. "Do you tink dis trip is a yoy ride?"

At night, however, with no stations to make, even Lindqvist and Karlson couldn't keep the men from relaxing. On many a night the moon raced in and out of clouds, the ship rose and fell in the glittering seas, and mixed with the sounds of the ship and the water came the melodies that Hanson Carroll was playing on his harmonica.

"Hey, Hanson, play 'Margie' again, ' one of the fo'c'sle gang would holler up the ventilator shaft, and Carroll, on lookout duty in the bow, would comply. Below, his friends would listen as they lay in their bunks or sat eating Italian cherries, the only canned goods that the raiding parties had yet been able to reach.

In the morning the routine would begin again, a station after breakfast for the scientists, more chipping and painting for the seamen. Dinner interrupted the work at noon and just as regularly as mealtime Arvid Karlson and Hans Cook met in the saloon at 2:30 P.M. for a game of cribbage. Neither man said a word during the entire game. On some afternoons they were joined by Lane, who liked to play solitaire on the gimbeled table while having a glass of wine.

Supper was served between five and six and shortly afterward Karlson would appear on deck with his sextant to look for the first star. Almost as invariably Backus's head would pop up through the midship companionway, for as chief engineer, chief ornithologist, and an expert on many subjects, he loved to try to beat Karlson at finding the evening star.

"Ver did dat Venoos vent?" mumbled Karlson while behind him the chief's head turned and twisted as he scanned the sky.

"Say, Karlson, take a pop at this one," Backus burst out, not noticing that the Swede already had his sextant to his eye. "It's been there for two whole minutes."

"Dat ting," growled Karlson, lowering his sextant in utter contempt, "has been dare for milluns und milluns of years."

With the sight finally made, Karlson lumbered aft to painstakingly work out the mathematics and enter the ship's position on the chart. He did not trust Loran — in areas where that service was available — and on previous cruises, when he had encountered Bumpus entering the hourly Loran fix on the chart, he had often smiled with kindly amusement and asked, "Ver doss dat fucking black box say ver ve iss now?"

Work continued even on Christmas Day, but on New Year's Eve the general routine was disrupted by a riotous party that swayed noisily back and forth between the after cabins and the deck lab. Bottles of rum taken on in Bermuda were pulled from lockers and

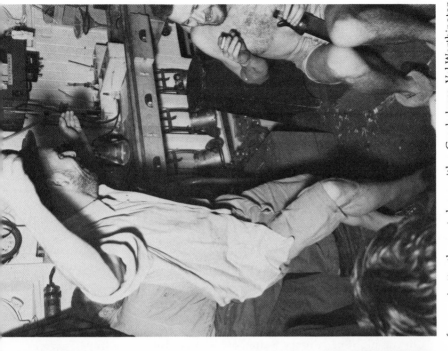

New Year's Eve party. Dean Bumpus happily drinks Sea Horse rum from a pipette, then cuts loose with a Greek dance. Val Worthington watches. (*David M. Owen photos. Courtesy Woods Hole Oceanographic Institution.*)

duffels, and at the height of the party — somewhat to ex–navy man Pollak's consternation — Bumpus was pouring pipettes full of Sea Horse rum alternately into Lane's beer and down his own throat. Bursts of laughter floated out on the quiet sea, and singing came now from the fo'c'sle and then from the captain's cabin, where Lane was playing his accordian.

As midnight approached, the ship's company began to gather around the pilothouse. A mild breeze from the south blew just enough to keep the sails well filled, and the ship rolled easily over the gentle swells. Men stood talking and laughing, some leaning against the bulwarks or hanging casually onto the rigging, a few checking their watches at frequent intervals. At midnight, according to custom, sixteen bells were struck, eight by Mattie Richard, a Nova Scotian sailor and the oldest man on the ship, and eight by the youngest, eighteen-year-old Albert Leonard.

No work was done on New Year's Day, except for twenty-three BT lowerings and the sailing of the ship, and the vessel seemed unusually quiet as she ran almost due east.

"The scientific staff has shaken down into a happy . . . group," wrote Dean Bumpus to Columbus Iselin. "We are looking forward to the next phase of the work."

On the night of January 5 the bow lookout raised the Cape Spartel Light, which sits on one of the northernmost promontories of Morocco, and early the following morning *Atlantis* sailed into Gibraltar. Five days later she was at sea again, rolling through the sparkling blue Mediterranean, bound via Malta for the famous Greek port of Piraeus, where her real work would begin. Within a day or two she left the fine Atlantic weather behind her and for a week and a half rolled and pitched through the violent winter storms that come howling across the Mediterranean from the north.

After a miserable buffeting the ketch stood into Salamis Bay and ceased her rolling and pitching in the lee of the rocky, gray-green hills that rise abruptly behind Piraeus. A pilot came aboard and guided the ship among a crowd of caiques, past outward-bound steamers, and alongside a huge, white-sided passenger liner into a welter of vessels that were tied stern to around a semicircular mole. An anchor was let go, the ship's stern was made fast, and like the victim of a pirate's raid the ketch was overrun with petty officials, wine merchants, tour guides, vendors, and pimps. A wiry old man jumped agilely aboard to sell genuine Greek antiquities from a dusty leather bag, and a stout woman in black used her four or five words of

English to solicit dirty laundry. The crowd became so thick that it was impossible to pass along the companionways, and the less English these determined salesmen spoke, the more loudly they shouted.

Finally, about suppertime, the ship quieted, and except for the crew under the watchful eyes of Karlson and Lindqvist, the men took their wads of paper money (12,000 drachmas to the dollar on the black market) and went out on the town. Bumpus and Pollak soon found that squid, octopus, and lamb guts were the local delicacies, and Val Worthington and Frank Mather (a naval architect turned BT observer) discovered that most Greek wines were too sweet for their taste. After tasting wines in several shops they eventually found one that struck them about right, and when they tried to tell the wine merchant that they were going back to their ship to get a five-gallon carboy, the man understood only that they were leaving and repeatedly lowered his already low price. The two returned in a taxi, bought the wine, and jounced back along narrow streets to the ship. Once the heavy carboy was safely aboard, they secured it under a desk in one of the scientists' cabins and fitted it with a device constructed of tubes, stoppers, and a squeeze bulb so that glasses of wine could be pumped out without moving the jug. Then back onto the mole they went, through the customs shed — which, limiting the night traffic onto the quay, protected the ships there from the forty thieves — and out again into the noisy, irregularly lit, strange-smelling, and altogether foreign port of Piraeus.

Piraeus was (and still is) very much a sailor's port. There were lots of bars, lots of brothels, and the people who worked in both had a sharp eye for a spender. They pegged the well-paid Dick Campbell as a heavy relaxer, and long before midnight Campbell had been shaken down. He returned to the ship for another pack of drachmas, and this in-again, out-again routine continued for as long as the ship was in port. Out went Campbell to one of several bars (he learned to leave his wallet and cigarette lighter on board after he'd had to buy back the lighter from a pawnshop), and several hours later back he came for more money. He had a lank, almost cadaverous build, and when he was drunk he was clumsy. The gangplank rattled more loudly under Campbell than under anyone else, and the deck seemed to amplify his uneven footfalls. But it was in the companionways, where he banged into a cabin door and ricocheted off the opposite bulkhead, that his passage was loudest. His processions became known as the "Campbell rattle," and for men in the after cabins who wanted to sleep they were a sore annoyance.

Toward the end of January *Atlantis* prepared to leave Piraeus to begin her work in the Aegean. The last of the stores were brought aboard about midmorning and the laundress returned with a single bundle containing all the shirts, socks, shorts, and pants of at least a dozen men. (It looked as if a hopeless sorting job lay ahead until the second mate noticed that she had sewn a thread of a distinctive color into the clothes belonging to each person.) Just after noon Harold Backus hurried through the customs shed with several more bottles of Mataxas. He found the others making two or three last grabs with the orange-peel sediment sampler right next to the ship in an attempt to retrieve Mattie Richard's false teeth, which had fallen out the night before when he had been sick over the side.

Finally Lane gave the order to weigh anchor, and while six men pumped the windlass, the anchor chain came rattling in. Suddenly the ship next to them began to move, and her sailors, shouting and waving their arms, indicated in some foreign tongue that their anchor was fouled with *Atlantis*'s. Lane steamed up on a magnificent snarl and after much maneuvering managed to clear the anchor and quit the inner harbor. He was now free to proceed to the fuel dock, which was the last step in the enormously complicated business of buying fuel. The fueling itself took a good part of the afternoon and evening, and it was nearly 8:00 P.M. before the ketch stood out of the Saronic Gulf. On either side rocky hills rose darkly and from points of light along the shore came the smells of charcoal fires and roasting meat. The silhouette of an old fort could barely be seen on a small island, and the bow lookout reported to Lane that someone in the fort was sending a blinker signal.

"I can't make it out," said Lane after watching a moment, his coat pulled close around him against the winter chill. "Go on ahead."

A second, much brighter light flashed near the first, and the sharp report of a rifle brought several men running up on deck.

"Hove to answering challenge shot across bow," wrote the mate in the log.

Still uncertain as to what was expected of him, Lane ordered half speed ahead until a second shot was fired. The ketch stopped again. Finally a corvette came out from the fort and once Lane had identified his command he was allowed to proceed.

The real work of the cruise now began, and *Atlantis* began weaving a complicated web of tracks among the islands of the Aegean. Down in the lower lab the echo sounder reflected the rough topography of the sea floor, while in the upper lab the ship's position was ascertained on the half hour. At night the radar was turned on and periodically Dave Owen would photograph the screen so that U.S. submarines might later locate their positions by matching the dots and blips on their own radar screens with those on one of Owen's photographs. In addition to the charting, two or three stations were made each day for water samples, temperature measurements, plankton hauls, and camera work.

Six days out of Piraeus *Atlantis* was occupying an anchor station near the island of Lemnos, in the north-central Aegean, to test two kinds of current meters. The winds, which had been light and variable, began to pick up, and within an unbelievably short time — not more than an hour and a half — a gale was sweeping over *Atlantis* from the northeast. As Lane ordered the engine on standby, the anchor cable parted and the ship was cast adrift to ride out the storm for the next twenty-four hours. Two days later a second gale caught the ship off Lesbos, and from February 8 through 16 the ketch encountered a series of violent storms that kept her bucking and pitching back into the lee of Chios, drifting out again into the steep, short seas, then standing in under the land again.

Atlantis eventually abandoned the protection afforded by Homer's island and made a run for Piraeus. She slammed along in half a gale until shortly before midnight on February 17, when she came rolling into Salamis Bay. Jogging on and off the breakwater outside the harbor all night, she stood in to Piraeus at 8:00 A.M. and in little more than an hour was again moored stern to at the Alexandros quay.

Because of the unusual number of storms, less work had been accomplished than planned, but since the main project was not one of the scientists' own choosing or one whose success or failure would greatly affect their careers, they were philosophic about their setbacks. Accustomed to Piraeus by now, several men immediately took the seven-cent streetcar up to Athens. Others stayed along the waterfront, and Lane went to negotiate with the ship's agent, who had "tried to slip us the splintery fid on the last bill."

That evening the third engineer, a good-looking young man with no allegiance to anyone but himself, who shall here be called Jay, got into what was becoming for him a characteristic state of trouble. Jay had already tried to smuggle a companion aboard in Bermuda, and had nearly been mobbed in Piraeus either for refusing to pay a prostitute or for trying to rob one. On this particular evening he was behaving comparatively well, and for several hours had been sitting in a cabaret, getting drunker and drunker. Val Worthington and one

"Out of three weeks we have had five and one half days of weather with sea state 6 to 7 and maybe even 9 for a couple of hours. . . . Nobody gets seasick, but it certainly makes you tired as hell." *(Conrad Neumann photo. Courtesy Woods Hole Oceanographic Institution.)*

of the sailors from the ship walked past the cabaret on their way back from Athens and spotted Jay slumped over a table near the door, his head rolling from side to side.

"I've been rolled, goddamn it," he mumbled as, one on either side, his shipmates raised him from his chair and moved him awkwardly through the door and down the cobbled street. The threesome wobbled down a long hill toward the water until the thought of the missing wallet reasserted itself in the engineer's foggy mind. Throwing himself free of his escort and cursing wildly, he ran recklessly, tripped over the first curb he encountered, and went somersaulting across the paving stones. He landed in a heap across the street, unconscious. Worthington and his friend carried Jay back to the ship and tried to lower him down the steep ladder that led through the main hatchway. Halfway down, the engineer's inert form slipped from their grip and fell heavily to the companionway below. He was carried the rest of the way to his bunk, and the following morning discovered he had broken his ankle.

"The crew is behaving better than the officers, on the whole," Lane wrote to Oakley. "The chief doesn't think too highly of [Jay], who has had in succession one street riot, one case lady fever, one eye infected, [and] one broken ankle."

"This fellow [Jay], my second assistant, is not very bright," Backus had already complained to Oakley. "He has absolutely no idea of electrical work. He did not know even how to test a fuse, and he can't use tools. . . . He has also been in a couple of scrapes here with the Greeks over women."

In the chief's estimation, the breaking of an ankle was by far the greatest of Jay's sins, for it meant that he would spend almost all the rest of the voyage hobbling around the ship in a cast — able to fish from the stern but not to stand his watch in the engineroom. It was back to six on and six off for Backus and Cook.

Lane was having no easy time in Piraeus himself. In addition to watching over his men, ordering parts for engines and instruments, and reprovisioning the ship, he had to try to obtain charts showing the locations of German mine fields planted during World War II and of hostile implacements of Greek Communists.

"This place is really running me ragged," he confided to Oakley. "The trouble is, the people who *can* speak English get you drunk by night and the people who can't heckle you by day. We sail in the morning, Zeus willing."

Zeus favored the sailing and *Atlantis* left for another three-week charting expedition, but, as Frank Mather, who knew a good deal about Greek history and mythology, pointed out, neither Poseidon nor Boreas concurred. In Greek, the work *cheimon* means both winter and storm.

In spite of the steep gray seas that rose like an endless succession of hills between the ship and the northern Aegean, *Atlantis* moved tirelessly on past snow-capped Mount Athos and up to the juncture of Greece and Turkey. From here she worked generally southward, charting the Gulf of Saros, heaving to under the lee of Lesbos, and often sailing close enough to the Turkish shore for her crew to see lights and navigation markers and to smell the orange blossoms that were beginning to bloom.

Atlantis continued south without mishap, and on March 12 Lane wrote with some satisfaction, "We have finished off the northern

Aegean, including Turkish waters (didn't get shot at either). I am certainly glad to get out of that damned hole. The weather this trip has been foul."

The next objective was to survey a portion of the Sea of Crete, which lies north of that island and south of the Cyclades, before returning for one last visit to Piraeus. Accordingly, the ketch began to work her way westward, and by March 13, as another cold gray dawn broke, she was well on her way across the Sea of Crete under shortened sail. To Lane, coming up on deck before breakfast, the lowering appearance of the northern sky and the bite of the wind indicated another storm. Even as he ordered the mizzen struck and a loose-footed trysail raised in its stead, the wind approached gale force and the seas made up into nine- and ten-foot piles of glistening foam that followed so closely upon each other that scarcely six seconds passed between successive crests. By breakfasttime the ship was rolling drunkenly along in the trough of the breaking seas, making better than eight knots under jumbo and mizzen trysail alone. Lane feared that the sails could not be lowered safely in the forty-knot winds that banged and rattled across the ship, and he decided that before heaving to, he would run for the lee of Cape Malea, one of the mountainous fingers of the Peloponnesian peninsula that extend toward Crete. But as the laboring ship approached the dark mountains that rose indistinctly through the clouds, the winds increased and the ship careened along more wildly still through a sea completely covered with foam. Lane had had no way of foreseeing that north winds grow stronger rather than weaker in the lee of the cape because they are funneled down long valleys. Since no other promontory or island offered shelter, the captain had no choice but to order the sails struck.

With a wind of sixty knots sweeping solid walls of spray and foam across the decks, and *Atlantis* often rolling forty-five and fifty degrees, well past the point where it is possible to stand, the crew was called out on deck. Karlson and several others grappled their way aft to strike the trysail. As they clung to the chartroom railing and the mizzenmast, the helmsman eased off the kicking wheel and let the ship come into the wind just enough to take the pressure off the sail that now racketed overhead. Engulfed in sheets of spray, Karlson and two others tried to ease off on the halyard, but the wind snatched the sail and easily ran out the line against their combined weights. With a wild ripping sound the trysail split and came crashing down, the block at its head narrowly missing Karlson.

Work on the jumbo, accomplished in three-foot waves that came flooding over the rail at every roll, was proceeding no better, and that heavy sail was also ripped before being brought in. After more than an hour's work, both pieces of sodden canvas were taken below. By this time *Atlantis* had been driven about a mile and a half offshore, and under full power she took two hours to recover the distance.

Belowdecks, resumption of the bridge tournament that had progressed during so many of these storms was not even suggested, and most men not on watch simply clung to their bunks to avoid being flung off their feet or seat.

Billy Cooper, a seaman, and one of the messmen, Albert Leonard, preferred to ride out the storm abovedecks, and they sat on the galley trunk hanging onto the half-open hatch. The ship, hove to again, was lying nearly broadside to the wind, and from where they sat, each glittering, foaming wave that came roaring toward the vessel as she rolled down to windward seemed about to crash its tons of green water straight down upon the decks. But at the last moment the ketch gave a lurch to leeward and, with her masts describing frantic arcs across the low sky, lay over on her other side and heaved herself sideways over the oncoming wave. A fifteen-foot wave picked her up in this manner and rolled her down fifty degrees, putting her lee rail out of sight underwater. The roll sent Cooper and Leonard flying off the galley trunk. As Cooper was catapulted into the main rigging, Leonard made a grab for the hatch, which shut on his hand, breaking one of his fingers and bruising the others. The two men scrambled below then, Leonard to get his hand bandaged, Cooper to join the crew at dinner.

Like Cooper and the rest of the crew, most of the officers and scientific staff had worked their way to the dinner table. If nothing else, a meal helped break the monotony of the storm. The men now sat clinging to the padded benches in the saloon, their knees held carefully askew to avoid the rhythmic passes made by the edge of the gimbeled table. Since Leonard was in too much pain from his broken finger to help serve dinner, the other messman was juggling plates of creamed salmon and peas on toast by himself. "Boston Blackie," as he was frequently called because of the resemblance he bore to a tough Hollywood private eye, was a small, wiry man who looked more like a jockey than a seaman. He was working hard to keep his balance, and slowly, as the ship allowed, he staggered back and forth from galley to saloon. If, on one of his spread-legged journeys into the saloon, he tripped over the lip of the hatchway, the captain and Karlson would get the salmon in their laps; if he set the plates down just as the table hung up on someone's knees, the food would slide across the unfiddled table and fetch up in the laps across the way. Blackie had managed to avoid both these accidents and had done no worse than pour hot cream sauce down his own arms until the ship gave a tremendous roll and the messman, rounding the foot of the table with a plate of creamed salmon in each hand, disappeared with a crash into the bosun's cabin.

By suppertime the storm had hardly abated, and although the cook and his assistants served another hot meal, as much from stubbornness as pride, many were simply too tired to eat. Night fell and the ship continued to alternate long periods of rolling in the trough with shorter ones of driving back into position against the wind. In his cabin Dave Owen lay wedged into his bunk, dispiritedly watching a light that danced wildly in the sky. It plunged into the flashing water, raced back and forth across his porthole, then shot into the sky beyond his line of sight. It was the moon.

During the night the storm slowly blew itself out, and by 8:00 A.M. wind and sea had abated enough for *Atlantis* to get under way and work slowly toward Piraeus.

"Got another breeze of wind yesterday from the NNE," wrote Lane laconically. "To hell with this part of the world!"

Atlantis made her final call to Piraeus on March 17, and this six-day visit differed from the previous two only in that Jay, who had been enjoying his new status as passenger, was still too lame to get into trouble, and that Lane bought himself a fourteen-foot, double-ended, lateen-rigged felucca from the British consul which he stowed on *Atlantis*.

The ketch was scheduled to sail from Piraeus early on the morning of March 23, but by breakfasttime the chief scientist, Dean Bumpus, felt so ill from something he had eaten the night before that he went over to the State of California's training ship, *Golden Bear*, to see her doctor. Roused from his sleep, the doctor had ungraciously dispensed several bottles of pills, and with these in hand Bumpus returned to his ship.

"Only an upset stomach," noted chief scientist Pollak in his log. "O.K. to sail."

After clearing the Bay of Salamis, *Atlantis* followed a characteristically circuitous track as she charted the sea floor southward through the Sea of Crete.

Dean Bumpus at work in the upper lab. Behind him Nansen bottles, each with a pair of reversing thermometers, are clamped in a rack. *(Courtesy Woods Hole Oceanographic Institution.)*

Bumpus, meanwhile, was passing several uneasy days. He tried to keep down his meals, or at least his pills, and he tried to keep up with the oxygen titrations that had to be performed on the water samples taken at each hydrographic station. Soon he became too sick for either food or work. While someone else took over the titrations, Bumpus simply lay in his bunk. His skin and even the whites of his eyes were turning yellow, Lane noticed with alarm, his pulse was weak, and his vomit was black. When he was no longer able to keep even a swallow of water down, Pollak suspended BT work "to avoid making winch noise over Bumpus's head," and Lane ordered the ship's course changed for the nearest port, Iraklion, in Crete.

Close to midnight on the following day *Atlantis* approached the ancient fortified harbor, lay to briefly at its entrance to pick up a pilot, then steamed in and moored in the lee of the north mole. The pilot immediately went to call a local doctor, and when the two returned in the dead of night they quickly climbed down into the chief scientist's cabin. There lay the jaundiced Bumpus, scarcely aware that the ship had come into port. The Greek physician felt his pulse, flashed a light into his mouth and eyes, then posed several questions to the pilot, who, as interpreter, passed them on to Bumpus.

"Ask him if he has a headache," the doctor must have said in Greek, and the pilot turned to Bumpus and asked, "How do you feel in *der Kopf?*"

Once the examination had been completed, the doctor produced a large bottle of pink medicine that, Bumpus later reported in surprise, tasted like a light Dubonnet. He kept the first dose down, and that, at least, was an improvement.

After breakfast the next morning, when it was clear that Bumpus was improving, his shipmates climbed up on deck in ones and twos to take a look at the port. It was a lovely harbor, its entrance protected by two moles built of light-brown rock and dominated in a mellow sort of way by a great stone fortress of the same material. The inner edge of the harbor was formed by the curving edge of the land itself, and beyond a fretwork of docks and wharves the low, light-colored buildings of the city sprawled across the sides of a dozen rolling hills. There were trees and larger expanses of green all through the city, and for the first time since entering the Mediterranean the men on *Atlantis* felt the unmistakable signs of spring.

"Jee-sus!" exclaimed Frank Mather, coming up through the deck lab and squinting through his wire-framed glasses at Homer's "island of a hundred cities." "Look at those hand-carved blocks of stone. Arabs took over this island in the ninth century and built that fortress there," he said to no one in particular. "Venetians came next, and later the Turks moved in. They beat the bejeesus out of most of the inhabitants, but it took them more than *twenty years* to take over Iraklion. Christ, you could have been born during the seige and still have had time to grow up and fight in it."

Mather told them too of powerful King Minos, son of Zeus and Europa, who had ruled Crete in mythological times, and of the extensive prehistoric ruins near Iraklion. In consequence, it was not long before most of the ship's company had strolled off down the broad mole to see what they could see either among the ornamented buildings and churches of the city or among the ruined columns and grassy excavations.

184

It was Easter Sunday, Lane noted in the back of his prayerbook, where he made infrequent entries of important dates, and although the holiday was not observed on that day by the Greek Orthodox inhabitants of Iraklion, it seemed a fitting Easter nonetheless. Bumpus was better, the day was clear and mild, and more than half the voyage was safely behind.

The weather did not hold, however. By late afternoon, when the first of the men were cheerfully and unsteadily returning to the ship, a twenty-knot wind was kicking up a steep chop in the outer harbor. Whitecaps were forming, and the larger breakers slapped into the mole and went hissing over its broad stone top. The men came slogging in through the water, and toward evening they were forced to hold onto the railway tracks that ran along the mole to keep from being washed off its back.

Lane and the dark-skinned pilot rowed out to *Atlantis* just at suppertime, saw the problem, and quickly agreed that the ship should be moved to the inner harbor in spite of the shoal depth there. The engine was started, the lines were cast off, and the ship moved cautiously in past the south mole. The pilot pointed out the most favorable ground, and at 6:20 P.M. both port and starboard anchors were let go in only eighteen and a half feet of water. *Atlantis* drew eighteen just then, and every ten minutes or so, when a particularly low trough passed beneath the ship, Lane felt the keel bump into the mud.

"I don't like it," he wrote Oakley that evening, waiting uneasily for the next thump, "but [I] don't dare move until morning for fear of doing worse."

The pilot agreed to stay the night in case the situation worsened, and that evening, and again before dawn, he climbed up on deck, unrolled his prayer rug, and, facing east, said his prayers to Allah. *Atlantis* came to no harm during the night, and before nine the next morning the anchors were brought home and the ship went on her way.

Describing triangles and irregular sawtoothed patterns along the northern and eastern shores of Crete, *Atlantis* continued to chart the sea floor until the Sea of Crete was left behind. Then, like a piece of wool drawn from a skein, her track straightened out and along its line three and even four hydrographic stations were made each day. The ketch sailed across the eastern Mediterranean in generally settled weather and put into Famagusta on Cyprus, her easternmost port of call. Her return trip through the Mediterranean was to be

Atlantis moored stern to in Valetta, Malta. *(David M. Owen photo. Courtesy Woods Hole Oceanographic Institution.)*

much more direct, with charting done only to fill in the gaps, or, as Pollak said, to "clear up the holidays."

After a two-week passage, the ketch again sailed into Valetta, Malta, and moored stern to at the old fish-market pier.

Officers and scientists had formed themselves into a solid group by now, and they knew each other's needs and preferences so well that there developed in each port a kind of scouting service in which everyone kept an eye out for a dry wine that would please Mather and Worthington or a new set of false teeth for Mattie Richard. There was a rescue service, too, an informal affair for the most part, but organized to some extent when Mather and Worthington decided to keep a running score of the number of men each was able to carry back to the ship. The competition had begun in Piraeus, and it had immediately become apparent that each of the ship's company would have to be rated on a sliding scale, for some collapsed every time they

Arvid Karlson, left, raises sail on *Caryn*. After working eight years on *Atlantis* keeping the "loofers" at work, he became master of *Caryn*. *(Jan Hahn photo.)*

left the ship on liberty while others, such as Chief Backus, had passed out only once ("It was them leekewers what done it, Val"), and one, Karlson, came in under his own power every time. He moved toward the ship like a dreadnought, Worthington remembered, with his companions weaving around him like destroyers.

It was not only for his ability to hold liquor that Karlson, like Ernest Siversen before him and a few men after him, became one of the legendary figures aboard *Atlantis*. He was not simply one of the ship's many characters, distinguished by the eccentricities that time at sea tends to exaggerate, but a man of integrity and unusual self-sufficiency. Few aboard the ship ever got to know the insular Arvid Karlson well, even after living with him for months at a stretch, but all respected him enough to wish that they had.

After clearing customs in Valetta, Captain Lane swung his felucca over the side and went sailing across the harbor while Bumpus and Pollak looked for unusual native foods. The crew, which in the old tradition conscientiously refrained from liquor at sea, went on a wild binge.

Lane had already learned that it was unwise to leave port after noon. Too many of his officers and seamen would be bumbling, cursing, falling-down drunk, their behavior too erratic to thread *Atlantis* through a crowded anchorage. Consequently, Lane tried to schedule his departures for the early morning, but in Malta this was not possible.

On the third of May, a mild, partly cloudy day with a gentle off-shore breeze blowing a strong smell of fish out over the harbor, a pilot came aboard at 2:30 P.M. ready to guide the ketch through a maze of fishing boats, freighters, tugs, and local traffic. To Lane's embarrassment, the man was forced to wait more than an hour for the last of the crew to be routed out of the bars. In dismay the captain watched the last of them come roaring down upon the ship. Singing and shouting, they wove their way among crowds on the pier, half carrying Boston Blackie, whose wiry frame twisted and writhed in repeated attempts to escape. Gassed to the hats, the men tumbled on board, and before Lane could ring up the engineroom, Blackie decided to throw himself overboard. He was halfway over the rail before someone grabbed him by the shirt and hauled him back. When ordered below, he took half a dozen steps in the right direction, then lunged for the side. Again and again he was pulled back, and for the first time in his command of *Atlantis*, Lane considered breaking out the manacles that were stowed in his cabin. He was on the point of ordering the messman tethered to a stanchion when Blackie was finally subdued and escorted below.

"You're worth a million dollars to me sober," Lane shouted in exasperation, turning on one of the mates who had accompanied the messman on his binge, "but when you're drunk, you aren't worth a goddamn cent!"

Lane stamped into the wheelhouse and ordered the engine started. He was up on the wheelhouse in a minute, and, regaining his outward calm, quietly directed his ship's departure.

"Needless to say, we are all rather impatient to get home again," he wrote Oakley. "It's been a long, tough trip and we're tired of it and of each other."

The charting project was finished, and rather like Prometheus carrying fire from heaven, Martin Pollak left the ship in Gibraltar to fly with a rough draft of the chart to the Hydrographic Office in Washington. Several of the other projects were all but complete too. Most of the thousands of BT slides, hundreds of water samples, and tens of bottom samples and deep-sea photographs had already been made or collected, plankton samples had been packed in wooden crates, and records of all kinds of phenomena, from wave heights to migratory birds, had been written down in a dozen notebooks. All this information would be worked up in the laboratory during the following year or years. Val Worthington and Nat Corwin eventually used the more than 20,000 miles of continuous soundings to construct a detailed bathymetric chart with contours at ten-fathom intervals, Bumpus and Bob Abel analyzed water samples and from their results generalized on the productivity of the Mediterranean, and Pollak interpreted physical data to ascertain the source of eastern Mediterranean bottom water. Of all the studies, this last stimulated new investigations of the Mediterranean and is generally considered the most important contribution.

On May 24, the day after Jay had been arrested for relieving himself on the wall of the governor's mansion, Atlantis cleared Gibraltar and within a day was rolling over the familiar swells of the Atlantic. Her mainsail, which in the crowded Aegean had rarely been used, was hoisted and set, and the ketch moved deliberately westward under full sail and power.

Seismic work, the only new study to be carried out on the homeward voyage, was begun the first day out of Gibraltar. The procedure was the one that Ewing had worked out on his first trip to the Mid-Atlantic Ridge, but the schedule was more relaxed.

For days a steady wind from the southwest sent the ketch rolling along toward Bermuda on a port tack, and every two hours a charge was hurled off the port side and a record made. The wind was gradually changing, however, and once when Bumpus came up on deck to stand his watch, he didn't notice that Lane had tacked ship and that the ketch was now sailing with her booms broad off on a starboard tack. Bumpus made up a charge, shouted, "Over!" threw it smartly off the port side, slightly forward of the wheelhouse. As the block and its bottle float left his hand, he caught sight of the mizzen extending obliquely before him and with undescribed emotion watched the twine connecting the float to the TNT wrap itself in a bollo hitch around the boom vang some three feet from the side of the ship.

"Clear the deck!" shouted Bumpus, running forward. To the men below there came a sudden, urgent thumping of running feet as the men tore forward and clattered through the hatchway. With a roar that shook the ship, the TNT exploded. Miraculously, the boom vang held, and the men, scrambling back on deck, could find no sign of the explosion. Gus Lindqvist was still crouched halfway down the ladder leading from wheelhouse to chartroom with one hand on the wheel. He had chosen to hang on, either from long habit or from the erroneous belief that the ship had to be kept on course at any cost, and although the bosun had not been hit by fragments of bottle or knocked down, his ears had been affected. The barometer, too, had been damaged by the shock wave. Knocked up to 31.80, it stayed there for the rest of the voyage.

The winds, such as they were, again came predominantly from the southwest, the sea became intensely blue, and at 7:00 A.M. on June 14 a lookout raised Mount Hill on Bermuda. With no motion wasted, Lane conned the ship into St. George's Harbour at 9:00 A.M., exchanged the necessary commodities in record time, and took her out again at 3:30 that afternoon. Under full sail and power *Atlantis* moved resolutely for Woods Hole. Uncovering a special screwdriver that he kept hidden for just such an event, Chief Backus readjusted the engine's governor and the ship moved ahead still another knot faster.

On June 16 there was one last burst of activity. For several hours the ketch maneuvered around an uncharted seamount and a camera station was made. (The deepest bottom photograph ever taken up to this time had been made earlier on the cruise at 3,026 fathoms.) Seismic work was accelerated.

As the ship approached the continental shelf, the sea became distinctly gray with a heavy tint of green and the open-ocean winds changed to summer breezes that were little help in speeding *Atlantis* home. Under power alone the vessel entered Vineyard Sound at dawn on June 18. As the sun began to rise into a bank of light haze

A saltwater shower on deck. *(Scott Bray photo.)*

above Martha's Vineyard, *Atlantis* was met by the Institution's dragger, *Asterias*. There was a lot of tooting and waving as the vessels drew together over the cool water. But Vineyard Sound was not Great Harbor and greeting parties were not wives and children, and so the ketch was soon on her way again up the sound.

On the Oceanographic's dock a crowd had already gathered when, shortly after 8:00 A.M., *Atlantis* slid silently into view from behind Nonamesset Island. To the crowd on the dock she seemed to enlarge rather than advance, and as the minutes passed, the details of her rigging, the line of men at her rail, and the familiar figure of Lane on the wheelhouse gradually became clearer. The ship moved in past the buoys at the harbor entrance, and the hum of her engine could be heard as she took a wide swing past the steamship authority and headed for the dock. As the ship approached, children grew suddenly shy. For a moment, those on the ship and those on the pier watched each other in silence, then someone waved, and forty arms went up. Still, both groups were reluctant to distract the crew during the precise maneuvers of docking, and it was not until the monkey fists had landed with a *whap* upon the dock and the first of the ship's lines had been slipped over a bollard that the half-shouted conversations mounted to an excited confabulation. The crowd moved toward the ship. On the sidelines stood a pregnant girl between her father and a policeman waiting for Jay. In another minute the sliding gangplank came banging down on the ship's rail, and *Atlantis* was once again a part of the land. The Med trip was over.

10 Pieces of the Puzzle Fit Together

It was [Maurice Ewing], more than any other man, who provided fuel for the revolution in earth science. I once asked him, "Where do you keep your ship?" He replied, "I keep my ship at sea." Over forty years he and his colleagues kept their ships at sea and provided the major part of our new knowledge.

— Sir Edward Bullard

In the first months of 1950, as *Atlantis* sailed southward on her annual winter voyage, Columbus Iselin decided to step down as director of the Oceanographic Institution. At forty-six, with a mild itch to "do science" again, Iselin was tired of administration. He had no time to work with the physical oceanographers who were planning an unprecedented six-ship survey of the Gulf Stream or the biologists who were trying to extend their work into the deep sea. Instead, he fenced with the dispensers of the navy's budget and listened to Maurice Ewing's pleas for ship time on *Atlantis*, "presented to the usual accompaniment of low moaning noises."

For a time there was "no definite dope about a successor," as Port Captain Oakley wrote Captain Lane, but as spring moved slowly northward, bringing birds and flowers and *Atlantis* back to Woods Hole, it was announced that Rear Admiral Edward "Iceberg" Smith would be the Institution's third director.

Eddie Smith, as few were allowed to call this rather formal gentleman, was retiring from the Coast Guard. He had begun his oceanographic career as a navigating officer in the International Ice Patrol (run by the Coast Guard) and had pursued formal studies in physical oceanography at Harvard under Henry Bigelow and later at the Geophysical Institute of Norway. In 1928 he had been chief scientist of the Coast Guard's ambitious *Marion* expedition and had explored the waters west of Greenland. For a short time Smith had commanded

193

Admiral Edward "Iceberg" Smith, left, the Institution's third director, stands with Columbus Iselin in front of the main laboratory building. *(Courtesy Woods Hole Oceanographic Institution.)*

194

the Coast Guard base at Woods Hole, but since World War II he and his family had been part of a much bigger installation on Staten Island. At sixty-one, this "kind and cautious man," as Iselin called him, was about to "swallow the anchor" and settle in Woods Hole.

Although Smith did not prove to be an active or inspiring leader, the choice of a military man — and one who was neither a friend nor a follower of Iselin's — was significant. It was an attempt on the part of the Institution's trustees to leave the Harvard Yacht Club image behind and put the Institution on a more businesslike footing. It was argued, probably correctly, that the Oceanographic would neither attract nor be able to handle increasing amounts of federal money unless it dropped some of its traditional informality and began to operate in a more conventional and presumably efficient manner. When, for example, the operation of the Institution's ships were described to Admiral Smith, he found the routine casual to the point of both danger and illegality. None of Iselin's old "I know you have your own ways of handling this" for Smith. He demanded periodic fire and boat drills, the issuance of permits for carrying explosives, and the posting of regulations. The Admiral found the dock area so disorderly that he instituted a personal tour of inspection every Friday afternoon to make sure that tools were put away and that the floor was swept.

Each of the new director's changes was met with ferocious opposition, in part because Smith did not have the personality to persuade; he simply regulated. As had happened to Samuel Clowser during the Institution's earliest days, this concern for superficial orderliness set him up as something of a scapegoat. More important, his desire for growth and professionalism ran counter to many employees' wishes to keep the Institution small.

In the spring of 1950, just before Smith and his family moved into an imposing house overlooking Little Harbor, the physical oceanographers at the Institution and their counterparts elsewhere launched the dream cruise they had been planning for well over a year. Operation Cabot was the project's formal name, and its aim was to use six research vessels simultaneously to make a detailed study of the Gulf Stream from Cape Hatteras to the tail of the Grand Banks. Under the direction of the Naval Hydrographic Office, whose leaders had noted laconically that "in the interest of efficient operations it becomes necessary to observe the behavior of the Gulf Stream," the operation was divided into three phases.

For the first phase, a general reconnaissance, the Royal Canadian

Navy vessel *New Liskeard* and the Bureau of Fisheries vessel *Albatross III* joined *Atlantis* and the Oceanographic's smaller ketch *Caryn* as they sailed southward to meet the converted seaplane tenders *San Pablo* and *Rehoboth,* lent by the U.S. Navy. In parallel all six ships zigzagged eastward with the stream, making BT slides and taking Loran fixes on the half hour and using the geoelectromagnetic kinetograph (the GEK, pronounced Geek), a newly perfected instrument for measuring surface currents. After five days the ships reversed direction and came back toward Hatteras.

From the thousands of BT slides thus made and the daily charts constructed, several things became evident. First — and this was no surprise to Fritz Fuglister — it was seen that as the Gulf Stream crosses the continental shelf off Hatteras, it begins to wobble and meander. Second, the "warm core" of the stream does not gradually decrease in temperature, but exists as gobs or pulses of warm water. Finally, instead of being a widening, weakening current, the stream seems to be composed of a series of narrow, overlapping streaks of swiftly moving water. In sum, the picture that emerged was so different from the traditional concept of a river that members of Operation Cabot suggested that the term "Gulf Stream" be changed to "Gulf Stream system."

During the second phase of the operation each of the six ships made a hydrographic section across the system. It was hoped that in this way a meander could be found to study during phase three. To the surprise of some and the pleasure of all, a suitable loop was located south of Halifax.

For the final ten days of the cruise the ships' tracks were determined by the movements of "Edgar the Eddy," as it moved some 160 miles to the south and finally detached itself entirely from the stream.

Several meander theories sprang up in the wake of Operation Cabot. It was suggested, for example, that the eddies might be analogous to the wavelike perturbations known to exist in the atmospheric jet stream, or that they might be caused by certain weather conditions or configurations of the sea floor.

"Obviously," wrote Iselin in reviewing the cruise, "physical oceanography has a long way to go before oceanic circulation is thoroughly understood. Nevertheless, having recently broken away from studies of the unreal, average ocean, we can expect rapid progress to continue during the next few years."

Late in June, just after *Atlantis* had completed Operation Cabot, Admiral Smith moved into the director's office at the west end of the laboratory building and inherited the administrative headaches that Iselin had left behind. Foremost among them was Maurice Ewing.

Ewing, since 1949 the director of the Lamont Geological Observatory, was trying to work at sea along two fronts. On the one hand he was interested in seismic studies that could reveal the deep structure of the continental shelves and ocean basins. For this work he ran seismic refraction lines with two ships, experimented with the less useful reflection techniques, and, as an indirect check on the first procedure, studied records from seismographs installed in a vault on the Lamont estate. At the same time Ewing was eager to study the more accessible marine sediments and the sea floor's topography. For all his work he needed a ship, and since *Atlantis* was the only research vessel on the east coast that could meet his needs, he was constantly banging on the Institution's door and complaining that WHOI and the navy (whose contracts paid for most of his ship time) were trying to limit his projects and strangle his research. What he needed for the summer of 1950, he told them both, was full use of *Atlantis* and *Caryn*. Largely because the Institution's physical oceanographers were saturated with data from Operation Cabot and its biologists and geologists lacked government contracts that would pay for ship time, Ewing was given what he asked for.

Thus, with Admiral Smith's consent, *Atlantis* left the Institution wharf on the morning of July 8 and moved out into Vineyard Sound as the sun burned through a light summer haze that hung over the Elizabeth Islands. The sound of her engine and even the voices of the crew came clearly back across the water as she hove to in order to accept explosives from a smaller vessel. Soon the peaceful summer day was intermittently disrupted by seismic reflection shots.

In addition to seismic work, which would eventually involve the ninety-nine-foot *Caryn*, Ewing desperately wanted to core sea-floor sediments. From cruises made to the Mid-Atlantic Ridge he had brought back several puzzling cores containing layers of well-sorted sand interbedded with the usual deep-sea oozes. It was a maxim of sorts at the time that sands and gravel were strictly continental shelf sediments and were not transported into the deep sea, where planktonic oozes covered all. But Ewing had found sand in deep water, and since the cores containing these anomalies had been taken from the broad plain that spreads out from the deeply submerged mouth of the Hudson Canyon, he wondered if the plain were not in fact a submarine delta that somehow received sediments from the continental shelf. To prove his hypothesis he needed more cores.

Coring operations duly commenced over the Hudson Canyon on

The ketch Caryn, built in 1927 and acquired by the Oceanographic in 1948, and the dragger Asterias, WHOI's very first vessel, both worked with Atlantis. Caryn was sold in 1958; Asterias is still in use. (Don Fay photo, left. Both Courtesy Woods Hole Oceanographic Institution.)

July 9. A heavy overcast had been blowing in from the southeast all morning, and by afternoon the whitecaps that slapped and sizzled along the vessel's hull shone with a curious iridescence under the leaden sky. As a piston corer (a device, modeled after a Swedish corer, in which a piston was used to help draw the sediment up the core barrel) was rigged amidships, the ketch was hove to, and to those who had not been on her since the year before, when her mainmast had been shortened twenty-eight feet to cut out dry rot, it was immediately apparent that the ship's rhythm had changed. Instead of rolling far over to either side with an almost ponderous gait, she now rocked back and forth with a snappier motion.

The corer made the first of many trips over the starboard rail at 2:00 P.M. and was safely back on board by 4:00. As the core was extruded on deck and taken into the lab for a cursory examination, the vessel got under way again and moved on down the canyon. Showers and squalls blew across the gray sea, casting shadows before them, and for the next eleven days *Atlantis* sailed through unsettled weather as the men on board took two or three cores each day. Gradually the ship approached Bermuda and just as gradually the dreary weather gave way to clear blue skies and a hot sun.

"Talk about hot!" wrote Lane upon arrival. "Ninety-one here in the cabin as I write this. . . . We have had quite a coring session — twenty-five to date. The wire is on its last legs, however, and I expect it to part every time we lower. . . . The wire is so bad [frayed] that it catches pieces of seaweed as it comes up."

To Lane's surprise, he found waiting for him in Bermuda a full set of Coast Guard regulations concerning the handling and stowage of explosives. Attached was a letter from Oakley informing him that "as matters now stand," by which he meant with Smith as director, "you will have to comply with these regulations, even though it means limiting the amounts to be carried or spending some time and money to construct said magazines."

"I shall check with the Coast Guard and naval authorities about the explosive loading," responded Lane. "I take it you want everything perfectly legal for once."

The next leg of the cruise, made in company with *Caryn*, was primarily concerned with seismic refraction profiles. One vessel hove to and listened while the other cruised away, periodically detonating explosive charges. Then the procedure was reversed, and the listening ship ran a line of shots as she caught up with the other vessel. Proceeding in this manner, like two ends of an inchworm, *Atlantis* and *Caryn* crept west from Bermuda across the sparkling summer sea.

The year before, Ewing had used *Atlantis* and *Caryn* to make a thirty-four-mile profile between New York and Bermuda, and for the first time he had brought back a fairly accurate idea of the structure of the ocean basin. Beneath several thousand feet of sediment the refracted

Ewing triumphant. Straddling a Steson-Hvorslev corer, Doc gazes rapturously at hunks of rock that have just come up in the dredge. Dave Ericson, left, holds one piece, Carl Hayes the other. Frank Press looks out to sea. *(Courtesy Dick Edwards.)*

sound had indicated a dense basement or oceanic crust. Unlike the continental crust, which was known to be twenty or more miles thick, the ocean's crust was only three miles thick and was composed of a much denser rock. The data strongly suggested that continents and ocean basins were two very different structures.

Now, as the two vessels sailed fitfully along, Ewing took cores whenever time allowed. Frayed trawl wire was first cut away, then inadvertently lost as the ships worked off Hatteras.

"We are now lying at anchor taking wire from *Caryn* with all the explosives aboard," wrote Lane from Norfolk on August 10. "The captain of the port said . . . that since the navy [rather than the Coast Guard] was loading us, we could load without a permit. Therefore I did not comply with the magazine requirements. . . . I hope you concur in this. The next time we load in Woods Hole I suppose we shall have to."

Lane was right. Several months later, after trying to regulate the stowage of explosives for half a year, Smith finally announced to Port Captain Oakley that *Atlantis* would not leave the dock again until fitted with wooden magazines. Although the ship was scheduled to depart the following day, and although Port Captain Oakley personally believed that if regulations were followed to the letter the explosives would have to be suspended forty feet up the ship's masts, he nevertheless cast about for a way to obtain six portable magazines in twenty-four hours. With characteristic ingenuity, he called the local furniture store, which doubled as a mortician's supply house, and ordered six pine boxes. The clerk was overcome with sympathy. He offered sincere condolences on so terrible a tragedy, and gently apprised Oakley of the fact that even those wishing a simple burial no longer used pine boxes. He could, however, let him have six papier-mâché copies of the more popular and luxurious metal coffins.

"No, no," groaned Oakley, and, quickly explaining his needs, was pleased to learn that he could get six wooden packing crates in which coffins had been shipped to the store. A perfect solution. Presumably to Smith's pleasure, six portable magazines sat on the deck of *Atlantis* as she set off the following day.

Smith was beginning to get his way. On subsequent cruises a special metal cap locker was lashed on top of the wheelhouse and chocks were made for the elliptical aerial bombs that before had been prone to shift as the ship rolled. Wooden magazines were carried for small amounts of explosives. Within the Institution itself such regulations gradually eroded the intimacy that had prevailed since its founding, but the changes were felt less on *Atlantis*. Paradoxically, her old-fashioned design and now her family-style operation were the very qualities that attracted men to her.

On Ewing's cruise in August of 1950, however, explosives were still casually piled on deck in haversacks as *Atlantis* and *Caryn* circled back to Bermuda to work in the oppressive heat that hung over the islands. *Caryn* lost her cook (he developed an "unbalanced mind"), and *Atlantis,* now loaded with an awesome array of 500-pound aerial bombs, got her stern line fouled in her propeller at the fuel dock. Using a diving helmet, Lane managed to chop away the line with a hatchet.

"This whole show has been a complete flail," wrote Lane, "and I am looking forward to the end of this trip with more than the usual enthusiasm."

There was still a month left to go, however, and as the sky took on a deeper autumn blue and the nights grew pleasantly chilly, *Atlantis* and *Caryn* headed toward Halifax. Taking roughly one core a day, and for the first time holding fire and boat drills, the vessels made their seismic profiles across rugged seamounts and flat abyssal plains. They called briefly at Halifax in early September and returned to Woods Hole through patchy fog toward the end of the month.

Ewing was well satisfied. From the Hudson Canyon area alone he had obtained ten cores, and all but one contained layers of sand or even gravel.

How did it get there? he had asked himself several years before when he had first encountered the anomaly. Now, having studied the sea floor for fifteen years, he felt ready to pull his data together and answer some of his own questions.

The picture that emerged from his work was one of a vast area — an ocean basin — that was fundamentally different from a continent in both structure and topography. Ewing sumised that the basins had existed for several billion years in their present form and were gradually being filled in by the remains of microscopic plankton, oozes, and continental detritus that was washed onto the continental shelves by rivers. A mechanism by which these continental shelf sediments could be transported into the deep sea had already been suggested in 1936 by a Harvard geologist, R. A. Daly, who proposed that, at least in times of vastly lowered sea level, huge density or turbidity currents might form like undersea avalanches at the shelf edge and race down the continental slope, carrying vast quantities of sediments with them. In an attempt to show that such avalanches could occur, a Dutch scientist had made miniature turbidity flows in a laboratory tank. His currents had spread sand onto his "sea floor," depositing the coarse grains first and carrying the silts far down the tank. Coupling these theoretical studies, which had not received much attention, with his own observations, Ewing now declared that turbidity currents offered the best explanation for the presence of the sands and gravels that he had found along canyon floors. In addition, he suggested that as the turbidity flows swept down the continental slope at great speed, they had cut and scoured the submarine canyons.

Still another piece of the puzzle seemed to fit into place when Ewing suggested that the enormous volumes of fine silts and clays carried by turbidity currents might slowly settle out of suspension far beyond the mouths of the canyons. Such a process would gradually fill in the irregular depressions of the sea floor and create the broad, flat abyssal plains that Ewing had in fact discovered seaward of the canyons. Almost all of the observations on which these ideas were based had been made on the cruises Ewing had undertaken on *Atlantis* in the late 1940s and in 1950.

During 1951 the ship's schedule changed hardly at all from what it had been for the past three or four years. As usual, Ewing used the ketch more than the men who were actually affiliated with Woods Hole, with the exception of Brackett Hersey and his group, who had been studying underwater acoustics since the late 1940s. Others at the Oceanographic Institution used *Atlantis* only occasionally, and some openly resented Lamont's extensive use of the ship. Ewing, for his part, resented any restriction of his projects, and since about 1949 he had been putting pressure on the navy to help him break free of the Institution by providing Lamont with a ship of its own.

11 Trade Wind Cruises

> He who would go to sea for pleasure would go to hell for a pastime.
> — Sailor's adage

To the dismay of Captain Adrian Lane and the crew of *Atlantis*, only four cruises were planned for 1952; together they constituted two exceptionally long voyages to the equatorial Atlantic. The decision to adopt this unusual schedule was motivated both by the Institution's desire to complete its general reconnaissance of the western North Atlantic and by the navy's need for sonar charts of the little-known region around the equator.

The first of the four cruises would take *Atlantis* to Dakar, Senegal, on the west coast of Africa, back across the Atlantic to Recife, Brazil, and finally up the coast of South America as far as the island of Trinidad. During the voyage studies would be made of the north equatorial current system and of an area of upwelling off northwest Africa. Much of the time *Atlantis* would work in concert with the 179-foot steamer *Albatross III*, which the Oceanographic had chartered from the Bureau of Fisheries.

On January 15, as high clouds moved across the sun, draining color from the sea and sky, *Atlantis* backed away from the pier and moved out into the cold wind that blew across Vineyard Sound. Still within sight of the shivering crowd on the dock, the ship hove to and loaded explosives from *Asterias*, then headed down the choppy sound. At first the wind blew with a cold exuberance, gathering the high gray clouds and flinging them in wild streamers across the sky, but then rain squalls hit and even the hardiest ducked belowdecks to be rattled along toward Bermuda. For four days the ketch pitched through an inauspiciously hostile sea, and even after she had paid a brief call at St. George in Bermuda, the heavy weather continued.

It was a grim crossing. Although the winds only rarely reached gale force, they blew with relentless monotony from the east. Gray seas built up, and *Atlantis* pitched incessantly. She slammed into waves that all but stopped her as fifty-gallon oil drums, packed with explosives, were wrestled over the side; she heaved and yawed in the black seas each midnight when a plankton tow was made; and she lay almost broadside to the wind and drifted smartly back toward Woods Hole as long strings of Nansen bottles were let over her side. There was no course, no speed, and no task that relieved the situation. She simply slogged away, the BT going down on the half hour, night and day.

As *Atlantis* worked doggedly eastward, once splitting her mainsail and only rarely using jumbo and mizzen, the far more powerful *Albatross III* set out to follow her. This narrow vessel, prone to rolling and unable to lift her bow over steep seas, had on board a crew of characters, as Lane would have said, and was skippered by the nervously alert Eugene Mysona. Shorty, as he was more often called, had joined the Institution as a messman aboard *Atlantis* in the 1930s and had worked his way up to the level of mate and relief master. Like a substitute teacher, he had more than his share of problems, and on this cruise he had scarcely departed for Bermuda when trouble erupted among the crew.

At the center of the scraps was Duke, a former oiler in the merchant service, a professional wrestler, a union sympathizer (in a nonunion crew), and "a real tough monkey." A heavy, belligerent man, he soon had the men on his watch thoroughly intimidated. Two jumped ship in Bermuda. They walked into the White Horse Tavern and wouldn't come out. Eventually the men on *Atlantis* learned by radio that Mysona had hired two Bermudians to fill out his crew and that the *Albatross III* was on her way to Dakar.

Meanwhile, Captain Lane on *Atlantis* realized that his ship would have to make an unscheduled stop in the Canary Islands for fuel — provided she could make it that far. Forced to motor against the wind, the ship was already low on oil although still ten days out of Las Palmas. Her speed was further reduced to conserve fuel and at about three knots she crawled through the first week in February against constant headwinds and contrary currents. Considering the hundreds of dollars a day that it cost to run the ship and the value of her scientists' time, even Lane, who was tremendously fond of *Atlantis,* could see that slogging across an ocean at a walking pace was not an effective way to do marine research.

On the cloudy afternoon of February 9, *Atlantis* finally rounded the light perched at the entrance of the Puerto La Luz on the island of

Las Palmas. Her fuel tanks were emptier than Lane had ever seen them before. For two days she remained in port, and while she took on fuel, her crew sauntered through the small gardens that bloomed in the squares and sat in the bars that opened onto tree-lined streets. Many of the men came back to the ship for dinner, and the calls of the newsboys, the ice cream vendors, the orange sellers, the fishmongers, and the scissors sharpeners drifted with them onto the massive mole.

Atlantis left Las Palmas on a flat sea and began describing a complicated pattern of zigs and zags that would eventually take her to Dakar. On this leg of the cruise a zone of upwelling, the result of constant offshore winds blowing the surface waters away from the coast and thereby allowing deep water, rich in nutrients, to well up to the surface in their place, was to be examined. The chemists on board, or more precisely the biologists-turned-chemists, were constantly at work, and both the upper and lower labs were crammed with their sample bottles, filtration racks, titrating equipment, and reagents. Each hydrographic station provided water samples, and each sample was tested for salinity, dissolved oxygen, phosphate, phosphorus, and so on.

The weather for the ten-day voyage to Dakar divided itself between familiar headwinds and nearly flat seas. In the galley a fine new steward, Joseph Lambert, made a little strawberry wine ("better than drinking vanilla," he claimed) and in the radio shack Sparks got word that wrestling matches were being held on *Albatross III* and that one messman was in bed with ulcers.

On February 21 Cap Manuel Light was sighted shortly after breakfast and by dinnertime the ketch had slipped past freighters, pilot boats, and the beautifully carved and painted dugout canoes of local fishermen to moor at mole 1 in the port of Dakar.

In spite of her late departure and delay in Bermuda, *Albatross III* was already in port and her crew already in the bars when *Atlantis* arrived. Soon the two crews were strolling along busy wharves and crowding into cabarets together, swapping tales. One of the seamen taken aboard *Albatross III* in Bermuda, a small, swarthy man, had been put on Duke the wrestler's watch and had received the usual bullying. Complaining to Captain Mysona had gained him no protection, nor had his reputation for making voodoo dolls, which he claimed to singe over a cigarette lighter and run through with a pin. Duke was immune to psychic blackmail, and the pushing and shoving had continued. Then one morning, when the two were on deck together, the Bermudian had found a flying fish. Picking it up, he viciously bit its head off, crunched the skull once or twice between his teeth, and swallowed. Duke was appalled. Seeing Duke blanche, his watchmate quickly ripped off several more bloody bites. Duke nearly got sick. From then on he referred to the Bermudian as a "goddamn cannibal," but he never bothered him again.

When not swapping stories or working to reprovision the two ships, the men from Woods Hole bartered shirts and shoes for bottles of cognac or went sightseeing in town. Some saw a French Foreign Legion parade with rank upon rank of jet-black Senegalese soldiers, others took their notebooks to the great sprawling fish market and recorded the varieties of local fish, and several sailed the ketch's

Eugene Mysona was captain of the *Albatross III* and later of *Bear*. (*Scott Bray photo. Courtesy Woods Hole Oceanographic Institution.*)

whaleboats out into the bay to circumnavigate a fortified island that had been used to imprison slaves bound for North America.

After five days in Dakar, *Atlantis* and *Albatross III* left for Recife, Brazil. The westward crossing was easier for both. For *Atlantis* there was good sailing weather in the trades, and for Captain Mysona there were no personnel problems, even though the galley staff was short-handed, as the messman with ulcers had been left in Dakar. On the ketch three hydrographic stations were generally made each day, a plankton net was let over the side at midnight, the bathythermograph was lowered regularly, and the GEK was used on the infrequent occasions when it worked.

Of all the data acquired, the water samples were proving the most useful, for it was becoming apparent that the equatorial current could best be traced on the basis of its unique chemical characteristics. Although diluted as it crossed the Atlantic, the current, with its particular chemical signature received in the region of upwelling off Africa, could be followed to within a few hundred miles of the South American coast. BT readings were less useful because this current system, unlike the Gulf Stream, did not differ markedly in temperature from surrounding waters.

After two weeks of fairly quiet routine, the ketch approached Recife, and on March 11, having passed between the two low moles that stretch along the city's vast waterfront, hove to to obtain pratique. Late that same hot and humid afternoon she moved into the inner harbor, past oil tanks, warehouses, grain elevators, and gawking cranes, and tied up outboard of *Albatross III*.

Seventeen officials, each perspiring in the uniform of a different agency and each expecting a handout, met *Atlantis* at the dock. Although Lane had been on the Med trip, he had never had to deal with organized bribery on so large a scale. He invited all the officials into the saloon, and over drinks shrewdly attempted to discern who had the power and who could be satisfied with a few packs of cigarettes.

Compared to Dakar — in fact, compared to almost any of the ports frequented by *Atlantis* — Recife was a big city. Fast-growing capital of the state of Pernambuco, it sprawled along either side of the Rio Capibaribe in the easternmost portion of Brazil. Into its huge railway yards, which extended under a smoky haze to the south of the port, came cotton, sugar, and other agricultural products; moving in the opposite direction went coal, oil, and wheat. All this commerce made Recife a busy, crowded city, but not a prosperous one. Even the most widely traveled seamen on *Atlantis* were stunned by the shabbiness, the poor health, the begging. In other ports the sailors were frequently heckled by groups of young boys who swooped down upon them like noisy crows, but in Recife, among its largely black and mulatto population, the begging was in earnest.

Another thing that amazed the men from *Atlantis*, as between rain showers they walked casually along the steaming streets, was the incredible amount of prostitution. Throughout the waterfront area, and for all they knew throughout the entire city, dozens upon dozens of coffee-colored women in cotton print dresses waved to passing men from balconies, older heavily made-up madams lounged along the sidewalks, and fourteen- and fifteen-year-olds worked the cafés.

A city ordinance prohibited the prostitutes from active solicitation before 6:00 P.M., but at the first stroke of the cathedral's chime they poured into the streets and the show was on.

The bosun on *Atlantis*, a hard worker but a red-faced, whiskey-winded drunk on the beach, took his pick of prostitutes without spending a cent. He had passed most of his spare time on the ketch trolling, and although happy to give most of the fish he caught to Frank Mather for Frank's study of large game fish, he would not part with the tuna. These he put in the ship's freezer for himself. Then, in Recife, just a few minutes after 6:00 P.M., the bosun hoisted a stiff tuna on either shoulder and marched off the ship.

"Where is he going?" Lane asked.

"Trades fish for flesh" was the reply, and sure enough, the bosun's shoulders had hardly been wet by the defrosting fish when he was surrounded by women.

On the following day, three of the ship's company, including Mather, who had sadly watched the tuna disappear, decided to have dinner ashore. They crossed *Albatross III* to gain the pier, and on an evil impulse each swiped a cooked lobster from a plateful left out for Captain Mysona. Hurrying off down a side street, they came to a café that opened out onto the sidewalk beneath a colonnade, and there they ordered wine and a few other things to accompany their lobster. From where they sat in the shade of the colonnade they could look out onto the street and watch the dock workers pass in their shabby work clothes and the barefoot children playing noisy games. They had almost finished eating when an old crone, dressed in black and looking, Mather said, like one of Macbeth's witches, shuffled in off the street. Hardly seeming to survey the cafe, she moved slowly toward the Americans. Silently she held out her hand, and without thinking one of the men popped an empty lobster carapace into it.

She was furious. Jabbing the lobster threateningly in the sailors' faces, the old woman screeched out a string of wild incantations. Waving her arms and gathering the forces at her command, she rolled her ill wishes together and flung a double, then triple curse upon the sailors and their ships. This done, her anger subsided abruptly and she shuffled back to the street.

Four days behind schedule because of the customs officials' refusal to authorize bunkering, both vessels left Recife, *Atlantis* with a change in scientific personnel, *Albatross* with a new messman. The same kinds of work were being done, and more than a week passed uneventfully before the curse of the old crone began to take effect.

On March 21 the ketch crossed the equator without ceremony and two days later was rocking northward across a monotonous succession of broad blue swells. The day was clear, with just a hint of a breeze, and after breakfast three or four men came up on deck to enjoy the morning. Gradually an intermittent clatter that had insinuated itself into the normal sounds of the pitching ship attracted their attention, and looking aloft they saw that the spring stay had broken free from the mizzen and was banging loosely against the mainmast. Without this forward support, the mizzenmast had already begun to sway, and with each easy dip and dive of the ship the mizzen leaned farther and farther aft.

"Frank! What'll I do?" shouted the young seaman at the helm, seeing the mizzen lurch toward the wheelhouse, where he was standing.

"I'd get the hell outta there!" yelled Mather, adding, as he ran for the deck lab, that of course he had no authority in these matters.

As the helmsman disappeared below, scrambling forward into the chartroom, there came to all belowdeck the painful sounds of cracking, splintering wood and shattering glass. The engine was immediately stopped, and as the ship settled in the water, her entire company rushed on deck. There, some twelve feet above them, stood all that remained of the mizzen. Its major portion dangled awkwardly over the port quarter, attached by a tangle of rigging. The boom lay broken across the wheelhouse, its two pieces held together by the furled sail.

Captain Lane, who had been in his cabin, where he would have been struck by the wheelhouse machinery had the spars come crashing through that sturdy structure, ordered the debris cut away. With fire axes and wire cutters, the sailors began to sever the rigging from the spars and let them slide into the sea.

"Partially dismasted at 04.44 north, 38.11 west," read the radiogram received with astonishment at the Institution that morning. "Mizzenmast, boom, and sail by the board. No one hurt. Program continues but may be somewhat modified."

After several hours' work, *Atlantis* (now the largest steel-hulled sloop in the world) continued on her way as before. The scientists began their work with buoys and a bathypitometer, a new instrument that measured water movement (that is, currents) versus depth much as the bathythermograph measured temperature versus depth.

For the next two weeks *Atlantis* worked northward, away from the

Shattered mizzen and boom lie across the wheelhouse. The mizzen toppled when the spring stay connecting it to the mainmast broke. *(Courtesy Woods Hole Oceanographic Institution.)*

westward-trending coast of Brazil and the Guianas, and on April 12 made her scheduled rendezvous with *Albatross III* in Port of Spain, Trinidad. Here the first trade wind cruise came to an end. Several scientists from both trips left for Woods Hole and several of the crew were fired.

"Require your reason for landing [messman]," radioed the port captain at the Institution to Mysona.

"Letter from Trinidad will explain," replied Mysona, who then discretely revealed that in addition to ulcers, inebriation, black magic, and dissention, he was now forced to deal with "perversion."

The next cruise, number 179, was under the direction of Maurice Ewing. It was not a part of the trade winds program, but instead was intended as an exploration of the structure of the central and western basins of the Caribbean. *Atlantis* and *Albatross III* were to take cores and make seismic profiles from Trinidad to Galveston, Texas, and finally back to Woods Hole.

"The plan is noted as both good science and of practical importance to naval defense," wrote Admiral Smith, cautiously straddling two potentially incompatible forms of science, and the program would have been carried out had not the Brazilian crone exacted her final retribution.

Trouble began three days out of Trinidad, when the trawl wire broke on *Atlantis*, quickly followed by the collapse of the corer itself. A replacement was requested from *Albatross*, but no corer was forthcoming. That vessel was steaming at full speed for Guantanamo Bay with Captain Mysona in desperate straits.

Earlier that same afternoon, April 20, a seaman with an unknown history of nervous breakdowns had come to Mysona to beg the captain to protect him. The crew was going to throw him overboard, he said. Mysona believed his story, and although he was uncomfortably aware of the lack of respect he commanded among his crew, he came down off the bridge to put an end to the bullying. The men, however, maintained their innocence, and when the sailor accused them of having smoked marihuana and having threatened him with knives, Mysona realized that he wasn't dealing with a rational complaint. Down in the uncomfortably warm mess, immersed in the residual smells of a hundred meals, the captain earnestly entreated his crew to assuage the seaman's fears. They tried to do so, and as Mysona went to get a phenobarbitol tablet for the man, they awkwardly told their shipmate that he was a regular fellow — likable, in fact — and that they would stick by him in any trouble.

Under the influence of the drug and perhaps the expressions of friendship, the seaman slept for a while after supper, but upon waking began talking so irrationally that Mysona ordered Duke the wrestler and his watchmate the cannibal to take turns watching the man at all times.

"[Seaman] has nervous breakdown," Mysona wired the Institution as he changed his ship's course toward Cuba. "En route Guantanamo Bay to land him."

The guard on the seaman changed with the watches, and as the quiet ship moved steadily through the night, her running lights hardly brighter than the tropical stars, his guardians reported periods of lucidity alternating with bouts of psychotic behavior. At breakfasttime the "patient," as he was now being called behind his back, seemed fairly calm and with some coaxing was led into the messroom. Expressionless, he asked permission to go the the head, but once free of his guards ran into the galley. Grabbing several carving knives, he dodged into the steward's empty cabin and locked himself in.

Mysona was immediately alerted. He joined the urgent discussion that proceeded in low tones outside the steward's cabin and agreed that the door must be jammed shut and barricaded so that the seaman could not get out and harm the others.

But what will he do to himself when he finds himself locked in? worried Mysona as he hurried along the companionway toward the radio shack. He had Sparks request medical advice from Guantanamo and was told, "Do not harass man. If man can be induced to come out by someone in whom he has confidence, give 8 cc intramuscular injection paraldehyde."

Mysona, remembering the sweat that had poured off him when he had first given a man a penicillin shot, almost grew ill at the thought of injecting 8 cc into a squirming, hysterical man.

"Leave man alone as he is not harmful to himself," came a slightly more reassuring bulletin, but Mysona was not comforted. He could not make port for another twenty-four hours and he was in torment over what might happen. Desperately he led small groups of men down to the steward's cabin, hoping each time that the seaman could be persuaded to come out. He could not. The afternoon sun slid slowly across the sky, the ship moved resolutely northward, but the seaman would not come out, nor would he throw the carving knives out through the cabin ventillator, nor would he eat or drink. Mysona had heartburn.

Late that night the captain retired to his cabin to write an account of the situation. As he sat at the desk, his ship rolling like a pig as she wallowed through a squall, he heard footsteps pounding toward him along the companionway. Without waiting to be summoned, Mysona burst from his cabin and followed the sailor below. The silent men gathered around the barricade were staring at the floor, and in the dim light the captain saw a trickle of blood sluggishly advancing and retreating from under the door as the ship rolled.

"Break down the door," Mysona whispered, and the barricade was stripped away and the metal door buckled inward.

To the captain's astonishment, out walked the seaman, smiling. His second finger lay on the floor; his little finger was hacked half through.

Applying a tourniquet, Mysona bandaged the man's fingers and with trembling hands gave him a shot of morphine. Again he commended him to the care of two crewmen, and soon they needed all their strength to hold the delirious patient in his narrow bunk. For hours he kicked and thrashed, then in the morning fell into a deep, coma-like sleep.

With the seaman in the custody of the navy at last, Mysona fired Duke and the bosun (the latter had bitten a man in the neck), said a cool good-bye to one of the scientists who refused to remain on this floating asylum a moment longer, and took his ship out to rejoin *Atlantis*. The two vessels finally began running seismic profiles. Within a day the program was suspended when the tail shaft on *Atlantis* cracked, as it had done on one of Ewing's Mid-Atlantic Ridge trips. This time Lane was not going to let the propeller fall off, at any rate. As Ewing rallied his group to take a core, swearing that he would not put up with such a bucket of bolts when Lamont got her own ship, the crewmen rigged a chain and manila preventer from the quarter break to the propeller. This done, *Albatross III* came alongside, and the biggest steel-hulled sloop in the world accepted a towline and was pulled toward Guantanamo Bay.

Repairs on *Atlantis* took almost three weeks, and it was not until mid-May that the ketch and *Albatross III* finally began the last leg of their long equatorial voyage. Ewing's plans, which had been trimmed drastically to fit the time available, called for a seismic profile each day and cores whenever the opportunity arose. The vessels steamed northward through the Bahamas, up the east coast, and finally through Vineyard Sound and into Woods Hole. On May 28, *Atlantis* nosed up to the pier, and in spite of Admiral Smith's attempts to save man-hours by discouraging the welcoming home of ships, a larger crowd than usual was on hand to see the sloop. The cat jumped off, the customs men walked on, and although Lane did not know it at the time, it was the last time that he would participate as captain in the ritual of return.

As his letters to Oakley had shown, Lane was getting tired of pushing the old rust bucket around the course.

"I can't help being annoyed at whoever made up the schedule, supposing so much work could be done in so short a time," he had written on an earlier cruise. "It's no wonder we have to put up with a crew of characters — no man in his right mind would stand it for long."

With the Institution's consent, it was decided that Lane would spend every other year ashore as port captain while John Pike would

Three captains of *Atlantis*. When Adrian Lane, right, became port captain, John Pike, center, took command of the ship. A year later Winfield Scott Bray became her master. *(Courtesy Woods Hole Oceanographic Institution.)*

alternate with him as master of *Atlantis* and port captain. The plan was actually put into effect, but although Lane stayed in Woods Hole for almost a year as port captain, sending his regards via Pike to "the Rat Pit," he did not hesitate long when the Marine Museum in Mystic, Connecticut, offered him command of the schooner *Brilliant*. Although Lane wrote that he was "not leaving Woods Hole through dissatisfaction, but in order to attend better to family affairs," he was finding that under the pressure of an expanding budget and the influence of Smith, the relaxed atmosphere at the Institution was gradually changing. On *Atlantis* the trend was underscored by the retirement of the last member of the ship's original crew, Chief Engineer Harold Backus, and by the departures of both Arvid Karlson, who left to take taciturn command of *Caryn*, and Gil Oakley, who, finding he could not work with Smith, had left the Institution

215

and turned the job of port captain over to the admiral himself. Lane was left with a less sympathetic group to work with, and the captain was not altogether displeased when, in the spring of 1953, he moved his family back to the quiet Connecticut shore.

Some months earlier, in July of 1952, *Atlantis* had left Woods Hole on the second half of the trade winds project. With a new master and a new and slightly shorter mizzenmast, she had sailed away on a four-month voyage that included both a further study of the Mid-Atlantic Ridge and a hydrographic survey of the Brazil current.

"It can now be said that so far as the North Atlantic is concerned, the reconnaissance phase has been concluded," wrote Smith when the ship returned, perhaps thinking of the 60,000 temperature measurements on record at the Institution. "Few sizable areas remain that are devoid of temperature and salinity observations."

In the admiral's opinion, the two long cruises marked "both the end of a long-term plan and the beginning of a new one. Having established a reasonably reliable picture of the average current system over the whole North Atlantic basin, we can now turn our attention to studies of departures from this mean pattern in more limited areas."

It would be a while, however, before physical oceanographers could digest the mass of data brought back from the equatorial Atlantic and proceed to detailed studies of limited areas. In the meantime, Brackett Hersey with his acoustical studies and Ewing with his geophysical ones continued to dominate the ship's schedule. Hersey and his colleagues took the ketch out in company with *Caryn* on a series of short, cold cruises in February of '53, and in April, when *Atlantis* was halfway through a shortened version of her customary winter voyage, Ewing's group from Lamont took the ship over in San Juan.

On April 21, 1953, which Captain Pike in his characteristically conscientious manner noted was a warm, clear Tuesday, *Atlantis* cast off from the tender pier in the port of San Juan and nosed her way through the cheerfully bobbing coconuts, wood chips, orange rinds, and blobs of oil toward the clearer waters outside. In true Lamont fashion the men aboard were already sweating over their gear, and the vessel had hardly cleared the sea buoy when the first explosive charge was thrown over the stern.

As on so many of these cruises, the main objective was to make seismic reflection and refraction profiles. This time most of the work would be done over the mountainous Sigsbee Knolls, which lie beneath the west central portion of the Gulf of Mexico. On the way, and to some extent throughout the cruise, coring, dredging, and the continued testing of the two new instruments would be done as well. One of the new instruments, a precision depth recorder, had initially been tried on the preceding cruise. The instrument consisted in part of a standard transducer, which instead of being mounted on the hull of *Atlantis*, where it would often rise and fall through a flurry of bubbles, was towed in a "fish" behind the ship. This procedure cleaned up the signal that bounced off the bottom, which was received as a trace of the sea floor aboard the ship, but did not solve several other problems that stood in the way of the development of a truly accurate system. A better recording device was needed, and men at Lamont had hit upon the idea of using a radio facsimile recorder (similar to that used to transmit photographs to newspapers) in conjunction with the towed transducer. The system was called the precision depth recorder, and although the recorder portion had already been taken off *Atlantis*, tests were still proceeding with the "fish."

On the first day out the bright-yellow, dolphin-shaped case was swung over the side and allowed to stream out behind and below *Atlantis*. As the steady trade winds blowing in from the northeast moderated the heat of the climbing sun, records of a steeply descending sea floor came into the lower lab. The ketch was heading west-northwest across the deep channels north of Hispaniola, and most of the day and on into the night the stylus continued to sweep across the slowly moving paper, drawing a dramatically exaggerated profile of the rough, steplike descent.

"Fine and clear. Rolling sluggishly," wrote Captain Pike near midnight, augmenting as he always did the succinct weather notations that marched down a narrow column in the log. His watch over, he gave the second mate the course to follow on the gyro and magnetic compasses and told him of recently sighted ships whose paths might cross *Atlantis*'s under the starry sky. Thus assuring himself that all was well with his command, the captain retired to his cabin.

John Pike was a first-generation seaman. There had been no nautical tradition in his family, no childhood filled with sailboats and maritime adventures, yet he had developed a liking for the sea and as a young man had elected to attend what in the 1930s was the Massachusetts Nautical School. Later he had joined the merchant marine and had spent several years aboard a ship carrying ammunition. Pike had come to Woods Hole in 1948 to inquire about a job with the

Bureau of Fisheries, and when told that the bureau's dream of a new research vessel was not likely to be realized for eight or ten years, he had walked back up Water Street and talked with Gil Oakley at the Oceanographic. By the end of that summer he was master of *Caryn*, and he stayed on her until he was transferred to *Atlantis*.

Pike was a calm, somewhat reserved man, a typically undemonstrative Yankee, and an excellent sailor. He could maneuver a sailing ship the way others managed a vessel under power, and more than once he had brought *Caryn* in under sail and "parked" her at a crowded dock in a slot hardly larger than the vessel herself. Proud of good seamanship, Pike wanted *Atlantis* kept shipshape and Bristol fashion, and in his zeal to run a tight ship he had of necessity become something of a bear for discipline. He was not, however, without humor, and was generally liked as well as respected.

Shortly after noon on April 25, *Atlantis* started a firing run. With charges hurtling off her stern, she steamed southward to close with the three-masted schooner *Vema*. The 202-foot iron-hulled vessel (built for the investment banker E. F. Hutton in the 1920s, in the same Danish shipyard as *Atlantis*) had been chartered by the Lamont Geological Observatory. When the navy had gone back on its agreement to furnish a ship, Ewing had taken an option to buy the racing yacht. If he acted upon it, Doc and his colleagues would leave *Atlantis* and thus change that ship's schedule dramatically. Such thoughts were not being considered, however, as *Atlantis* moved sedately across long, sparkling swells to approach the impressive vessel.

Lying hove to, her sails for the most part dumped on deck, *Vema* rolled easily on the blue water. From her graceful concave stem to her stern the schooner's lines were elegant. Even without her topmasts, which had been removed some years before, her masts rose 120 feet above the deck. She carried a thirty-foot bowsprit, beneath which, framed by the ship's dolphin striker and lit by light scattered upward from the waves, rode a fierce golden eagle. *Vema* gave the impression of energy and speed: she had established the unofficial record of ten days and ten hours for a transatlantic passage, and she seemed the kind of ship Ewing would use. By evening the schooner had become the shot boat, and leaving *Atlantis* silently hove to, she sailed off toward the southern edge of a violently colorful sunset.

For the next three weeks the two sailing ships moved slowly across the Gulf of Mexico, and although Ewing complained that *Atlantis* couldn't keep up, that the ketch's rigging was in disrepair, and that as a listening ship she was severely handicapped by her limited battery

The Lamont Geological Observatory's first ship, *Vema*, was an iron-hulled schooner built in the same shipyard as *Atlantis*. Her masts were gradually cut down until nothing now remains of her rigging. She is still used as a research vessel. *(Don Fay photo. Courtesy Woods Hole Oceanographic Institution.)*

power, it was the schooner, not the ketch, that was reduced to sail power when her cylinder heads cracked. The loss of an engine didn't stop Ewing from running profiles, however, and by mid-May work over the Sigsbee Knolls had been completed. The vessels parted company in Galveston, Texas. When *Atlantis* left port, clearing the bar and turning toward home, Pike ordered all sail set for the first time on the cruise. Coring, and occasionally experimenting with a sediment thermometer that could be attached to a core barrel and sunk into the sea floor, the men on *Atlantis* worked eastward, then northward. The waters they sailed through changed from brilliant blue to gray and green, and the season retreated from summer back to early spring. On June 10, under a cold and cloudy sky, the ketch stood up Vineyard Sound, doused sail, and motored cautiously into Great Harbor.

"End of cruise 185," wrote Pike.

This cruise, at least modestly successful in all respects, was the last for Captain Pike, who became port captain, and the last for Ewing and his colleagues at Lamont. Although Columbia University was extremely reluctant to become a shipowner, Ewing manage to purchase *Vema* in a typically irregular manner only hours before his option expired. The character of *Atlantis's* schedule changed immediately: instead of making only two or three cruises in the second half of 1953, she made ten, and all for scientists at Woods Hole. Brackett Hersey's underwater acoustics program accounted for fully half the cruises, but there was time too for plankton hauls, experimental work with buoys, topographic surveys of newly discovered seamounts, and a short drift in the Gulf Stream. Superficially, it was as if the routine of the 1930s had been revived, with scientists of each discipline taking the ship in turn. In 1953, however, there was a suggestion that such short cruises were less an expression of interdisciplinary cooperation than a result of the limitations of *Atlantis.*

"The experience of the past six months indicates that *Atlantis* is too small to carry out a complete oceanographic program with modern instrumentation," Martin Pollak had written upon his return from the Mediterranean in 1948 (the same year that the Scripps Institution of Oceanography, on the West Coast, had acquired two sizable ships from the navy). Even Iselin, with his emotional attachment to the vessel, had admitted in his last director's report that although "*Atlantis* is still the flagship . . . , the continued advances in oceanographic techniques and field equipment are causing cramped quarters and restricted space."

To some it seemed shocking that the twenty-two-year-old *Atlantis* was no longer ideally suited to meet modern demands, and they blamed her designers for a lack of foresight. In part they were right. Even before 1930, sailing ships had been well on their way to becoming more of a luxury for those with time and money than a true economy. The ketch had been something of an anachronism from the start. On the other hand, Bigelow, Iselin, and others had had no way of predicting World War II and the extensive changes that oceanography would undergo as a result. *Atlantis* had been designed to serve a small summertime laboratory supported by a modest endowment, not an institution handling large projects paid for by the navy.

But whether the ketch's limitations came as a surprise or not, there was a problem, and Iselin responded by accepting a contract from the Office of Naval Research to study the requirements of modern research vessels and to suggest several new designs. His report was completed in 1953, and within its long list of recommendations could be read a request for a new vessel, a replacement for *Atlantis*. A fully powered ship was needed, one longer, beamier, heavier, faster, more comfortable, and better protected from the weather.

Since the Office of Naval Research was already committed to financing a new laboratory building for the Institution and was also beginning to support an open-ended study of the hydrography of the western Atlantic, it was not ready to provide a ship as well, and the request was not met. And so *Atlantis* continued to be the Institution's flagship and major research vessel. Even though the scientist who used her most, Brackett Hersey, said that her lack of power and of heavy lifting gear had struck him the first day he had boarded her in 1946, the ketch was still the largest privately operated research vessel available to him. With the exception of the eight or ten naval ships assigned to support the surveys and applied research of naval laboratories, only *Vema* at Lamont and *Horizon* and *Spencer Baird* at Scripps were larger.

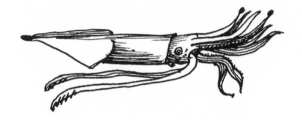

12 Underwater Acoustics

Contract B [from the navy's Bureau of Ships] was the broadest contract that the Oceanographic Institution ever received. It said, in essence, "determine the effect of the ocean on underwater sound."
— Allyn Vine

To anyone working at the Woods Hole Oceanographic Institution in the early 1950s, it was obvious that the group of scientists studying underwater acoustics under the direction of J. Brackett Hersey was an anomaly. The most striking distinction was the group's degree of autonomy. While the rest of the Institution's staff worked together in an atmosphere frequently described as "pleasant chaos," Hersey's people charged across the frontiers of acoustic technique and instrumentation as an organized and independent unit. Furthermore, his group was larger, richer, and in some opinions more highly motivated than any of the others.

The reasons for these differences were complex and were based on the group's extensive involvement with classified research, its financial independence, and so forth. But the result was fairly simple. "Hersey's gang" operated like an institution within the Institution. Hersey himself had done what Columbus Iselin wouldn't do and what Admiral Smith, in many cases, couldn't do: he had put together a unified group.

Brackett Hersey had come to Woods Hole in 1946. He was a geophysicist and, like his former professor, Maurice Ewing, an intensely energetic man. Hersey's long-term objective was to study the complex relationships between underwater sound and the things that can affect it, such as the physical properties of seawater itself, the surface and floor of the sea, and animals, which can either mask the sound with their grunts and cackles or scatter it with their bodies. He was also interested in perfecting the hydrophones and other equipment used to receive and record underwater sounds and their echoes.

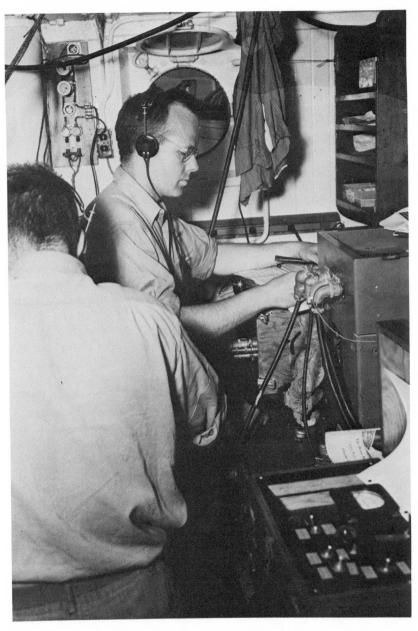

J. Brackett Hersey tests electronic gear in the upper lab on *Atlantis*. (*Don Fay photo. Courtesy Woods Hole Oceanographic Institution.*)

On acoustic cruises concerned primarily with "active listening," Hersey loaded *Atlantis* and *Caryn* with explosives and bounced sound off the sea floor, off tame submarines, and even off a newly discovered layer of animals that in many parts of the ocean rose toward the surface each evening and descended at dawn. On other cruises he was more interested in "passive listening," and on these cruises he recorded the background noises of the sea — the sounds of rain, breaking waves, fish, snapping shrimp, and passing ships. In either case the navy was willing to support the work, for as it became increasingly convinced that the advantage in undersea warfare would lie with the side with the more acute hearing, it invested more and more money in underwater acoustics.

In the summer and fall of 1953, while *Atlantis* was making more than a dozen short cruises about equally divided between underwater acoustics and the other business of the Institution, Hersey was putting plans together for a long winter voyage to the West Indies. For two and a half months he and his group intended to use both *Atlantis* and the Oceanographic's smaller vessel *Bear* to make sound transmission runs, topographic surveys, seismic refraction profiles, trawl hauls, and a variety of classified studies.

On the morning of January 18, *Atlantis* and the Plywood Palace, as *Bear* with her high wooden superstructure was often called, left the Institution's pier, and to the usual accompaniment of shouts and waves headed into Vineyard Sound. A gentle north wind pushed them slowly across the blue water as they hove to to take on explosives from the work boat *Playmate*. By noon the loading was complete, and with a final exchange of good wishes *Atlantis* and *Bear* set out for the waters off Cape Hatteras, where the sound-transmission runs would begin. As the sky clouded over late that afternoon and the wind backed into the southwest, slowing the ketch down, *Bear* gradually pulled ahead. By nightfall her rolling silhouette had disappeared.

Bear, a useful but uncomfortable boat, was a 100-foot wooden vessel that the Institution had first used in 1951, then purchased in '52. Although she was faster than *Atlantis* under most conditions and had more room in her lab for electronic gear, *Bear* was not a handy vessel in any other respect. For one thing, she was prone to almost uncontrollable rolling, and her captain, Shorty Mysona (who had commanded *Albatross III* when she was chartered from the Bureau of Fisheries), frequently feared to turn her around lest she capsize when broadside to the seas. For another, she was abominably crowded and could not carry enough fresh water for either washing or bathing. She was noisy, too, starting off at low speed with a *hotchaka-hotchaka-*

hotchaka and increasing to a banging, clattering roar. In her galley the table was so arranged as to shoot catsup, pickles, and tableware out of the fiddles and pile them on the deck whenever the ship rolled heavily. Not surprisingly, the vessel's crew consisted of new hands unfamiliar with the ship's idiosyncrasies and eccentrics who for their own reasons were willing to put up with them.

By noon on January 20, both *Bear* and *Atlantis* were rolling along through a fleet of fishermen off Chesapeake Bay and the first of a long series of end-to-end seismic profiles was scheduled to begin. In a rising sea and spitting rain *Bear* came alongside to exchange last-minute instructions, then was quickly away. The noisy *pocketta-pocketta-pocketta* of her engine came back across the waves, and soon charges of TNT were flying off her wallowing stern. *Atlantis* rolled silently in the swells, all power shut down. On the ketch's deck hydrophone slackers in oilskins and seaboots hauled in the weighted hydrophones, and on the shouted command "Slack!" paid them out again. After an hour the roles were reversed, and *Atlantis* fired explosives as she came plowing through the heavy seas toward *Bear*.

Several of these profiles were run during the next two days and seven hydrographic stations occupied before work was stopped by a storm that blew the ships far to the south. After running several more profiles off Florida, they put into Miami.

On February 6, a mild, clear day in southern Florida, *Atlantis* left Miami a day ahead of *Bear* in order to make some hauls with the relatively new Isaacs-Kidd midwater trawl. This awkward piece of equipment had been designed by two men on the West Coast in the interest of capturing the animals of the middle depths, some of which were suspected of being responsible for the phenomenon known as the deep scattering layer. This migrating layer, first reported during World War II, was in many places so thick, or rather so reflective of sound, that it had sometimes been mistaken for an uncharted shoal when records of its shadowy form had appeared on echo sounders. Since the war the layer had been studied acoustically, and by the mid-1950s biologists were eager to discover what animals were migrating up and down in the middle depths. Although the new trawl could not be depended upon to catch a representative sample of the midwater population and thus could not conclusively answer the question of what squid, fish, or shrimplike animals made up the deep scattering layer, it could give biologists some idea of the possibilities.

As *Atlantis* moved into the Straits of Florida, the crew broke out the trawl and began the complicated job of rigging it for towing.

About 4:00 P.M. the ketch slowed until she was barely moving over the brilliant blue water. The trawl was swung out over the rail. It bumped disconcertingly against the side of the ship, then splashed into the water and rapidly sank out of sight. For the next hour the ship crept along through the warm afternoon at about three knots. At dinnertime the order was given to start the winch and bring the net in.

Dick Backus, a young biologist who had performed a multitude of tasks for Hersey — he was not related to the former engineer Harold Backus — was on deck as the trawl came swimming up through the clear water.

"Sight!" he shouted to the man at the winch controls, and the latter carefully brought the first section of the trawl's bridle to the top of the A-frame at the side of the ship. The net, with its unusual plow-shaped depressor, which caused it to tow deeper than an ordinary plankton net on a given length of wire, was still lying in the water, and it was necessary to rig a tackle from high in the ship's rigging to bring it in. With this second line secured, the trawl was brought banging up the side of the ship. Once the depressor was clear of the rail, it was swung inboard and set on the deck. Several men then grabbed the net, shook it as it came over the rail so as to knock any fish into the cod end, and pulled net and liner onto the deck. They dumped the catch into a plastic tub that had long since replaced the wooden buckets of earlier cruises.

To Dick Backus's disappointment, it was a scant haul: a few angler fish, the young of several species of flatfish, an eel, and some eel larvae. The entire catch could be scooped up with one hand, as eventually it was, and dropped into a mason jar for preservation.

The trawl was set again that night and twice on the following day. Slightly larger catches were coming in, and Backus was sorry to see *Bear* come huffing over the long swells to resume the seismic work.

On February 11, Hersey, who had joined *Bear* in Miami, and who made a practice of riding the less comfortable ship, transferred to *Atlantis* so that he could observe the operation of several pieces of experimental equipment. "Bud's Clunker," for example, was a mechanical pinger that was being tried as an alternative to an electronic model. It made a noise, all right, Hersey noted in his diary, but he could not distinguish it from the crackling and popping of snapping shrimp — sounds that abounded in the area.

On the twelfth the full complement of underwater acoustic work was under way, and Hersey, surrounded in the upper lab by a half-dozen men all fussing with recording devices and within earshot of the slackers on deck tending the hydrophones, was pleased. This, he thought, is the way a cruise ought to be run — every piece of equipment operating every minute of the day. At ten that morning sounds that were probably made by whales were picked up, and for the next two hours the helmsman on *Atlantis* was directed to steer various courses at various speeds as the ship chased after the ticks and groans.

At 6:00 P.M. *Bear* and *Atlantis* hove to within hailing distance, and Captain Mysona came bobbing over from *Bear* in a building sea to discuss several problems with Hersey. One concerned an eye irritation that was bothering *Bear*'s engineer, Bill Shields.

Whispering Willy Shields was a heavy-set man with a voice like a foghorn, two deaf ears, and a supreme insensitivity. His rude, often accurate observations, delivered with the coarsest profanities, endeared him to some of his shipmates but made others uncomfortable, among them Mysona.

Shields was a native of Gloucester, and since putting to sea at the age of thirteen he had had a legendary series of adventures. He had rolled 360 degrees in a fishing schooner (and had thus been spattered with hot lubricating oil, which caused his eye problems), received the Carnegie Lifesaving Medal for pulling a man from a burning engineroom, used his heavily muscled and tattooed arms to clean up on innumerable men in innumerable bars, and boasted of having to blow his tubes in every port. Shields did not remember the periods of his life by the houses he had lived in or the women he had lived with, but by the ships he had been on. He had been engineer on *Bear* before the ship came to Woods Hole, and rather than leave her, he had simply changed employers and stayed aboard. He sweated away happily in her infernally noisy engineroom, and kept her in such good tune that at slow and half speeds "ya don't see no smoke." As far as Shields was concerned, he was the final authority on the engine: no matter what Captain Mysona asked for on the engineroom telegraph, Shields gave him what he thought he ought to have.

In addition to the engineer's inflamed eye, Mysona had a problem with *Bear*'s Loran set, which had "crapped out again," forcing him to rely on celestial fixes.

Hersey promised to return to *Bear* on the following day to see what he could do for the Loran. As for Shields, he would be hospitalized in San Juan if his eye did not improve.

On February 13, as *Bear* circled around and around *Atlantis*, where two recording devices were being compared, the wind, which had been slapping whitecaps across the waters east of the Bahamas, began to blow in earnest. By noon it was obvious that Hersey could not be returned to *Bear*, and by evening it was so rough that the echo sounder was the only piece of scientific equipment still in use on *Atlantis*. To Hersey's intense disappointment, the wind raged in from the east for an entire week. On the nineteenth he finally gave in to the weather and ordered the ships to seek shelter in the palm-lined harbor of Charlotte Amalie on St. Thomas.

Three days later the weather cleared, the ships prepared to leave, and, as usual, Bill Shields and the assistant engineer dropped belowdecks on *Bear* to warm up her engine. But as Whispering Willy began moving the levers on the switchboard panel there was a sudden flash of fire. Shields leaped backward, the hair on his arm singed by the flame. Grabbing the hose on a carbon dioxide extinguisher, he directed a stream of gas and another of smothering profanities at the flaming board. Smoke and fumes filled the room, leaked upward to the main deck, where the vessel's explosives were stored, and came spiraling out of every port and companionway.

"Fire! Goddammit, fire!" bellowed Shields as he and the second engineer erupted on deck from opposite ends of the engineroom. The words weren't out of his mouth before every person on and around the ship was charging toward the crates of TNT that sat securely lashed beneath their tarps on deck. Lines were untied in a moment, covers thrown back, and the wooden boxes hustled off *Bear*. In less than two minutes, the last of the pile had been removed to the far end of the dock. The flickering flames that had risen partway up the ladder toward the laboratory and the smoke that had billowed briefly from the engineroom subsided of their own accord, and within half an hour navy volunteers with gas masks were able to go through the ship to extinguish the last sparks and turn off all electric switches.

There had been a moment's anxiety when two of the ship's company could not be found, but once Captain Mysona arrived and the ship's cook was located in his bunk, where he had been sleeping the entire time, it was clear that no one had been overcome by fumes. *Bear*, however, had sustained considerable damage to her electrical system. Plans were quickly made to tow her to a shipyard in San Juan, and late that afternoon *Atlantis* left St. Thomas for a week of unscheduled trawling.

This was a bonus for Dick Backus, and he lost no time getting the Isaacs-Kidd trawl into the water. By evening he was experimenting with a shallow tow at a depth of about seventy feet. The trawl was

Bear in Great Harbor. The 100-foot-long wooden vessel had such a high superstructure that she was nicknamed "the Plywood Palace." *(Courtesy Woods Hole Oceanographic Institution.)*

not brought in until after midnight, and as the net was hauled on board, every taut line shaking off showers of sparkling drops, it was immediately apparent that the catch was a good one. Out of the cod end of the net and into the tub slid a slippery stream of shiny black and red and silver fish.

Just like a horse shitting Christmas-tree ornaments, thought Backus with a smile, remembering Terry Keogh's legendary description.

Much of the catch was made up of young dolphinfish that had been caught as the net was drawn to the surface, and when the larger of them were cut open, triggerfish, squid, and unidentifiable remains of other fish were found in their stomachs.

Trawling was continued over the next five days, and from depths of 1,500 to 1,800 feet came large numbers of bright-red shrimp, lanternfish, a few pale jellyfish, grotesquely shaped angler fish, and some deep-sea species with colored light organs dangling from their barbels or set in arcane patterns upon their heads. Several dozen squid were caught too, three of which proved to be species never before discovered. One of them, given a Latin name meaning "sparkling moon squid," was a diminutive creature less than two inches long with light organs, large and small, profusely distributed along its body, around its eyes, and over several of its arms and tentacles.

All the animals were preserved in a Formalin solution and packed in mason jars, which for safety's sake were crated and lashed atop the upper lab.

Once *Bear* was repaired and Hersey's plans were reinstated, the two vessels worked with submarines on precisely choreographed maneuvers known as "events." There were Event Able, Event Baker, Event Charlie, and so on. During Event Easy, for example, *Bear* sat listening with hydrophones as a submarine near her went around and around in a large circle. Beyond the sub, *Atlantis* ran back and forth along a straight course, dropping charges at varying depths. Although the navy was loath to disclose its precise objectives in sponsoring these events, all were ultimately aimed at improving a ship's ability to detect enemy shipping while avoiding detection itself.

When the navy work was finished, *Bear* and *Atlantis* headed home. Seismic profiles were again run off the Carolinas, and it was only when the explosives ran out on March 30 that work was ended on the cruise. Three days later the vessels ran up Vineyard Sound and nosed into their accustomed berths along the Institution's pier.

For the next two months Hersey's group was busier than ever. Not only were there tapes and wiggle traces, fish and echo-sounding records, cruise logs and masses of other data from the cruise to analyze and report upon, but there was also a move to be made. From the cramped offices and laboratories scattered through the main brick building, the underwater acoustics group was moving fifty feet up the street into the new Laboratory of Oceanography, which had been built with funds from the Office of Naval Research. The intent was to bring together all the Institution's military projects, in part to make the protection of classified material easier. Hersey's group moved several tons of books and papers into the first and second floors, and physical oceanographers set up their BT files and other paraphernalia on the third. In June a colloquium was held to honor both the new building and the continuation of naval support that its construction implied.

A year before, in the summer of 1953, John Pike had left *Atlantis* to become port captain on a more or less permanent basis and Winfield

Scott Bray had become the ninth master of the ketch. Scott Bray was a handsome, solidly built man of forty who had divided his not altogether happy years among an unusual combination of pursuits — seamanship, art, and cooking. As a teenager he had left his home in Indiana to join the navy, and after serving for four years aboard a battleship had worked as a seaman on merchant ships and as a cook in a San Francisco restaurant. Upon his return to Indianapolis he had seriously studied drawing and painting.

At the age of twenty-five Bray married, and with his wife continued to experiment with various jobs and life-styles until World War II broke out and he enrolled in a U.S. Maritime Service officers' school. He graduated just a few days after the birth of his second child and almost immediately shipped out as an ensign in the merchant service. Bray stayed at sea for the better part of five years, and as he traveled in and out of ports all over the world, he took an interest in every culture and cuisine he encountered.

"He was the perfect tourist," one of his *Atlantis* friends said later. "He was polite and well informed. He could relax in style."

When Bray came ashore in 1949, more handsome and charming than ever, he held few memories in common with his wife and two children, and shortly thereafter the family decided to separate for a year. Bray and his daughter moved to Cape Cod. There he opened a restaurant, but, victim of a dishonest partner, he closed it a year later.

Bray was stranded, and as he had done at seventeen and again at thirty-one, he went to sea. Although he had more than a dozen years of seagoing experience and an unlimited master's license when he applied for a position at the Oceanographic Institution, he accepted a job as seaman aboard *Caryn*. Several months later, in July of 1953, he was given command of *Atlantis*.

Captain Bray had made only two short cruises on *Atlantis*, which could accurately be described as routine, before he encountered the minor calamities so typical of a research vessel's career. On the third cruise, the ketch had scarcely left Woods Hole on a warm August day when a seaman was brought before him and accused of erratic and aggressive behavior. Interpreting the man's threats to jump overboard, swim home, and kill his wife as the effects of too much liquor, Bray sent him forward to sleep it off. *Atlantis* continued down Vineyard Sound peaceably enough for an hour or so; then Bray heard shouts forward of the deck lab. Leaving the wheelhouse, he saw the seaman, armed with a long knife, clamber over the ship's rail and splash into the sea. The ketch was quickly brought about and Bray ordered a whaleboat lowered over the side. For the next five or ten minutes the seaman refused to be rescued. When the bosun leaned from the whaleboat to grab his arm, the seaman swung viciously at him with the butcher knife. After this had happened several times, the bosun decided that to save him he would first have to knock him unconscious with an oar.

"I don't want to die anymore," the seaman suddenly shouted as the bosun slapped the water around his head with a heavy wooden oar.

Repeating the phrase over and over, the seaman dropped the knife into the water and, catching a line thrown to him, let himself be pulled aboard.

"Put under custody unshackled," read the ship's log. "Back to Woods Hole."

One month later, Captain Bray again took *Atlantis* down Vineyard Sound, this time laden with twenty tons of oceanographic equipment that was being carried to Bermuda for a two-year project on wind driven currents. Her passage was uneventful, and in St. George Harbour the equipment was put ashore. The weather seemed unusually cloudy and unsettled to Bray, so he was not greatly surprised when, on the morning of September 17, the Coast Guard issued a tropical storm warning. Edna, they said, would move across Bermuda that evening with winds of 60 to 70 miles per hour. Although Bray was told that *Atlantis* would be safe at the dock, he took the precaution of moving his new command out into the harbor and shackling the vessel to F mooring. (This was a star mooring, held in place by five or more chains and anchors deployed in a sunburst or star pattern.)

By the time the ketch was made fast, the wind had already begun to build from the southeast, and soon it was whistling over St. David's Island, pelting *Atlantis* with sand, leaves, and pine needles. By dinnertime it was howling, and streaks of foam rose from the harbor like smoke. As night fell, the hazy silhouettes of other vessels tugging and pulling at their moorings and the larger shape of *Lord Cochrane,* a dredging barge that sat solidly in place, her buckets dug into the harbor floor, disappeared from view. At 8:15 Bray ordered the engine ahead slow to ease the strain on the port anchor chain, shackled to the mooring, and as gusts of wind came booming across the ship like cannon fire, Bray called for half speed, then fast half.

At 9:00 P.M. all hands were ordered to emergency stations, and the watertight doors throughout the ship were dogged down.

"Wind SE 100–120 mph," wrote Bray. "Two men at helm." Suddenly the quivering and shaking, all that could be sensed of the vessel's attachment to her mooring, ceased, and with a wild swing the vessel was off across the harbor, broadside to the wind.

"Let go the starboard anchor!" Bray bellowed into the wind. "Full speed ahead!"

Several men struggled forward almost on hands and knees and somehow managed to drop the starboard anchor. *Atlantis* swung back into the wind, but continued to move backward across the foaming, hissing water. She swung around her anchor again — perhaps it was holding — then dragged slowly in a slightly different direction, straight toward the blurred lights of *Lord Cochrane.*

"Emergency full speed," ordered Bray, and as *Atlantis* continued to drift, he rang the ship's alarm and passed the muffled word to stand by for a collision. Gamely the seamen on *Atlantis* unshipped the vessel's fenders and docking lines, and in the shrieking wind that stuffed each word they uttered back into their mouths and battered them with almost enough force to shove them overboard, they prepared to board the dredge and make the ketch fast to her. The lights on the dredge came closer, and her outline appeared, wreathed in spray. And then, as the men clinging to the rigging watched, she seemed to move away. The wind shifted a point or two, and the ketch, swinging clear, continued her backward circuit of the harbor.

With the immediate danger of collision over, an attempt was made to rig the port anchor on the parted chain. First Mate Dick Colburn, a down-Mainer whom Bray had found "no end of help in getting me acquainted with the problems I am likely to meet," crawled forward through the abrasive blast of sand and water that had already scoured the paint off the ship's ventilators and with several others finally succeeded in rigging the anchor and letting it go.

"Dragging both anchors. Wind shifting to westward, gusts 90 to 100," wrote Bray as *Atlantis* moved blindly toward the mouth of the harbor. A mooring buoy materialized out of the night and bumped roughly along the port side of the ketch. Surely, thought Bray, the anchors would foul on its chains. But the wind shifted again and the ketch circled toward St. George.

"Stone quay thirty feet astern, sir," reported the lookout stationed on the vessel's stern. It was almost midnight. The ship's engine was still on emergency full, straining against the wind, both anchors were out, and there was nothing more the exhausted crew could do.

"Wind shifting," wrote Bray a moment later in a scrawl so tired that the words fell off the line.

Slowly *Atlantis* changed course once again, moved parallel to the quay, then proceeded out into the harbor.

"Anchors fetched up. Wind dropping," the captain finally wrote half an hour later.

"I have heard many times how terrible the sound is that comes from the wind roaring and shrieking through the rigging," Bray wrote to his wife several days later. "Now I have heard it . . . And believe me I hope I never run into winds of such force again ashore or afloat."

Bray remained on *Atlantis* for Brackett Hersey's long voyage to the West Indies, and although he quickly learned that acoustic cruises were tiring, hard-driving affairs, he found himself more and more interested in all forms of scientific work and more and more fond of *Atlantis* herself.

During the summer and fall of 1954 *Atlantis* made more than a dozen short cruises, as she had done the year before. Two were to see if a deep-sea lobster and crab fishery could be developed off Massachusetts, several were for Hersey's group, two were for geologists, one was for physical oceanographers, another was for a project concerning the stereo photography of waves, and the last was to the Gulf of Venezuela for scientists of almost every discipline.

Wedged into this busy schedule was the triumphant installation of a power windlass to raise the anchors and two fresh-water showers, the vessel's first facilities for bathing.

To jump ahead for a minute, still another long Hersey cruise took off for the Caribbean in March of 1956. As usual, large quantities of explosives were loaded aboard *Atlantis* and *Bear*, and for almost two months the vessels worked in and out of the naval base at Guantanamo Bay on a dozen or more acoustic projects. By mid-May the men on both ships were exhausted, but seismic profiles continued to be made with relentless regularity on the homeward trip. As had happened many times before, Hersey's intense desire to study the transition zone between continent and ocean basin was working contrary to Mysona's desire to get *Bear* home. The two began to argue.

"Hersey," muttered Mysona, "is another Ewing, both like horses with goddamn blinders on," and once, when the captain found a man up to his knees in water at the stern of *Bear*, clinging to the ship with one hand and trying to light a fuse on a demolition block with the other, he was so incensed that he countermanded Hersey's instructions and ordered the work stopped. Even the ordinarily enthusiastic

It took six or eight men a lot of time and a lot of sweat to weigh anchor on *Atlantis*. Each complete up-and-down motion brought in one link of the anchor chain, or about six inches. *(Courtesy Woods Hole Oceanographic Institution.)*

Bray felt that "on Hersey's trips there are an awful lot of long days."

On the evening of May 16 the two vessels finished a profile well offshore, came about, and headed for Charleston, South Carolina. It was a point of pride with Bill Shields that *Bear* should always precede *Atlantis* into port, and knowing this, Captain Bray quickly ordered *Atlantis*'s engine full ahead and some sail raised in hopes of getting Whispering Willy's goat. He did, and within minutes Shields retaliated and *Bear* came churning past *Atlantis* at full speed. Three or four potatoes (surrogate grenades) came thumping down on the ketch's deck, and before *Bear* pulled out of range they were returned.

The two ships had a long way to go before reaching Charleston, and as night fell they had yet to cross the Gulf Stream, which ran between them and the Carolina shore. The breeze that had been blowing intermittently from the northwest picked up as the night wore on, and, on impulse, Bray ordered all sail raised on *Atlantis*. With her engine still going at full speed and the wind abeam, the ketch rolled down until her lee rail was awash and pulled for shore. She bounded across the choppy seas of the Gulf Stream, slid with a hiss and a roar through smoother coastal waters, and at dawn drew within sight of the low islands that lie to either side of Charleston's harbor. Far ahead was *Bear*, hardly more than a speck on the water.

At breakfasttime *Bear* had to heave to for a few minutes to bring her towed echo sounder back on board, and as she sat in the water *Atlantis* gained rapidly. At 8:31 *Bear* was moving again. Captain Mysona rang full ahead, and with a clattering roar the vessel settled her dark stern into the water and raced for the sea buoy. *Atlantis* gained more slowly now, but with her sails full and drawing, on she came. Everyone save her engineers was passing excitedly up and down her steeply canted deck, and potatoes were being gathered from the skiff that served as the ship's potato locker. Past the sea buoy and in toward the long twin jetties raced *Bear*, and right behind her now and slightly to windward pressed *Atlantis*.

"We've got you this time, Mysona!" shouted Bray into the radio as his ship began to come abreast of *Bear*.

"More power, Willy!" yelled Mysona, pumping the handle of the engineroom telegraph for all he was worth. "More power!"

But Whispering Willy was not in his engineroom. Having left the throttle wide open, he had gone out on the stern of *Bear*, and there he was doing his damnedest to beat back *Atlantis* by hurling potatoes across her bow.

"Weep, Shields!" radioed Captain Bray, and *Atlantis* pulled ahead and sailed majestically between the outstretched arms of the jetties.

Shields did indeed weep that evening, and for the four days that the vessels were in port no one let him forget his disgrace.

A return match was inevitable. The two ships left Charleston, and after completing the last of the seismic profiles they prepared to race for Woods Hole. Again a half-dozen eggs and potatoes flew through the air, and a scientist who kindly threw a package of pipe tobacco over to *Bear* had it mistaken for a potato and hurled back — almost. It became "ocean shag."

The two ships did not stay together on this run. *Atlantis* tacked out into the Gulf Stream to take advantage of the current while *Bear* made a noisy beeline for Woods Hole. All night long the two were out of sight of each other and all the next day.

"Hey, there she is!" shouted lookouts on *Bear* at irregular intervals, hoping to trick Willy into squeezing more power from his engine. Shields already had her going full ahead, and *Bear* rattled on until, on the second morning, at the edge of Block Island Sound, her engine coughed, gasped for lack of fuel, and died.

Mysona couldn't believe it. With his ears ringing in the uncommon quiet and his command bobbing serenely over hazy gray seas, he and several others hastily made plans to drain the dregs from all the fuel tanks into a single reservoir. Before they could start, however, the second engineer came on deck and explained somewhat sheepishly that before he had gone off watch he had forgotten to open the valves in the tanks that allow a reserve supply of fuel to be tapped. With these valves opened, the engine roared to life and *Bear* charged ahead into a dense bank of fog. That was no problem for Mysona, and with the ship's horn tooting and her radar running, *Bear* made all possible speed for Woods Hole. Later, as she entered Vineyard Sound, Captain Mysona saw a blip on his radar screen well astern of his own vessel. Getting on the radio, he tried to raise *Atlantis*.

"Is that you, Bray? I can see you on my radar." But Bray would not give him the satisfaction of a reply. The wind had died when the fog came rolling in, and *Atlantis*, her furled sails dripping and her engine pushing her along at an infuriating eight knots, came trailing into Great Harbor an hour after the triumphant *Bear*.

No more races, Bray and Mysona agreed, thinking especially of the first one. Shields would see *Bear*'s engine blown apart before he'd lose again.

In Brackett Hersey's opinion, both ships had lost the race. Both were too slow. For ten years he had worked with *Atlantis* and *Caryn* (the smaller ketch was now up for sale) and with *Atlantis* and *Bear*, and neither combination suited his purposes. Moreover, he believed that the liberal contracts he received each year from the navy would soon be diverted to naval laboratories and to the Scripps Institution of Oceanography in California, where much larger ships could gather data in far less time. Hersey resolved to leave Woods Hole if he could not get himself a larger ship. While Admiral Smith wrote memos and reports advocating the construction of modern oceanographic research vessels, Hersey and another of Ewing's former students, Allyn Vine, began to hunt in earnest for a castoff naval vessel with which to replace *Atlantis*.

13 Into the Pacific

I thought I heard the first mate say,
   leave her, Johnny, leave her,
"Just one more drag and then belay,"
   it's time for us to leave her.

— Chantey

The Atlantis Marine Geological Expedition to Chile and Peru began on the warm, partly cloudy morning of October 29, 1955. *Atlantis* and her crew were in Bermuda, where they had just completed a three-week study of waves, and although the ship was now leaving St. George to sail through the Panama Canal and enter the Pacific Ocean for the first time in her career, there was little gaiety at her departure. A small crowd had gathered along dock 4 to say goodbye, but because the expedition's scientists were not on board, because Captain Bray was not feeling well, and most of all because *Atlantis* carried the ashes of her former first mate, Arvid Karlson, the leavetaking was subdued. A restlessness rippled among the men like wind across a lake, and it was a relief to everyone when shortly before noon the ketch steamed out of the harbor and passed through the town cut. All that day and part of the next the ship drove southward across a gently ruffled sea until on Sunday morning orders came to stop the engine, back the jumbo, and heave to.

"Last mortal remains of Captain Karlson were put to eternal rest upon the sea in the presence of the entire crew," wrote Bray as the ship drifted quietly across the long swells.

Arvid Karlson. Bray had served under the big, slow-moving Swede on *Caryn*, and most of the men on *Atlantis* had been Karlson's shipmates on one or the other of the two ketches. Without closing his eyes Bray could see Karlson and his shipmates raising *Caryn*'s sails, the men squinting upward through the sunlight, their arms intertwined and their legs bumping as they crouched and stretched, crouched and stretched, pulling on the halyard. For Hans Cook, still chief engineer on *Atlantis,* the overwhelming memory of Karlson was of playing cribbage with him. Silently the two had sat across from each other in the saloon almost every afternoon for the eight years that Karlson had been on *Atlantis.* Others remembered Arvid Karlson shaking the vessel through a squall that had caught her with all sail up, or maintaining order in the wild ports of Malta and Piraeus on the Med trip, or with a broom sweeping the "gotdam whores" off *Caryn* in Martinique. Henry Stetson had once drawn Karlson's attention to a rotten spot on *Atlantis's* rail, someone recalled.

"I thought teak was supposed to last forever," he had said without much thought.

"Efen your fucking rocks don't last foreffer," Karlson had replied.

And now, at sixty-four, Karlson was dead, and the first business of the Pacific expedition was to honor his request and scatter his ashes on the sea.

The short service over, Hans Cook disappeared below, and the ship's engine started with a vigorous roar. Within an hour the sun came out, a breeze sprang up, and cruise 221 was under way.

The primary purpose in organizing this long voyage to the southeastern Pacific, a region studied infrequently by undersea cable companies and only once by scientists from the Scripps Institution, was to collect sediments from deep trenches that lie fifty to one hundred miles off the coasts of Peru and Chile. An interconnected series of trenches or "deeps" extends roughly from the equator to the north-central region of Chile, and Henry Stetson, perceiving the cruise to be the last major geological investigation he would undertake before retirement, intended to make a dozen profiles across this unusual feature. On half the sections, cores six to seven feet long, he hoped, would be taken with a Stetson-Hvorslev gravity corer, and on all the runs the ship's echo sounder would be in continuous operation.

For the six or eight geologists who would participate in relays in this program there were three main areas of interest. One was the shape of the trenches themselves, and what this shape might suggest about the time and method of their formation. Another was the sediments that lay in the trenches. The third concern was the possibility that the trenches were oil fields in the making — early stages of fields like those that exist on land along the north coast of Peru. This last question had prompted the Esso Research Lab and the Socony Vacuum Corporation to contribute funds to the expedition.

For a full week *Atlantis* steamed south and west under a hot sun,

Atlantis in the Panama Canal. The ketch had ascended the locks at the eastern end of the canal, sailed through Gatun Lake, and was about to descend to the Pacific. *(Courtesy Woods Hole Oceanographic Institution.)*

her passage boosted by a light breeze. As had happened only rarely since World War II, the ship was deadheading, that is, traveling without collecting data. There were no scientists aboard.

"It is almost lonely this leg without our usual load of scientists back aft here," wrote Bray.

On November 5 the thickly forested hills of Panama appeared along the western horizon, and as the sun set behind their dark shapes, *Atlantis* passed into Limón Bay and proceeded to an anchorage off Cristóbal. Late into the warm evening the crew strolled along the deck, watching the lights that sparkled in the city and extended in uncertain paths across the water toward the ketch.

Early the next morning, with the temperature already climbing into the eighties in spite of a thin cover of clouds, a pilot came aboard, and under his direction *Atlantis* steamed to the head of the bay. Here, between low hills of luxuriant green, she was taken in tow by two small switching engines or "mules" that ran along tracks on either side of the locks. With violent jerks the mules tugged *Atlantis* through the first gate and into a narrow enclosure. Sheer cement walls rose ahead and to either side of the ship, and a control station perched far above like a guard tower above a prison yard. The gate swung slowly shut behind the vessel, and moments later water boiled into the lock through unseen pipes. In less than an hour *Atlantis* had been pulled through the second gate and cast loose to sail across the rippling waters of Gatun Lake. By afternoon the vessel was dropping toward the Pacific, first through the Pedro Miguel Locks and then through those at Miraflores. At 4:04 P.M. the last gate swung open and Bray radioed to Woods Hole, "We have invaded Scripps' Ocean without firing a shot!"

For two hot, hazy days *Atlantis* sat stolidly at pier 18 in Balboa, where the Panama Agencies Company provided wharfage, removal of garbage, and the rental of a rat guard for $100.10. Fuel and water were pumped onto the vessel, stores were replenished, and two young geologists who had been visiting the lushly overgrown ruins of Old Panama came aboard. Bray, meanwhile, contemplating the lonely, unlit coast of South America in his mind's eye, took the precaution of getting a medical checkup. As in Bermuda, where he had been taken to a hospital after suffering a "seizure" in mid-voyage, and in Massachusetts, where he had later sought a second medical opinion, the captain was half relieved and half upset to learn that the pains in his legs and chest could not be traced to any abnormality. With the most common yet consummately useless advice for a worried man — "Relax!" — he was sent back to his ship.

On November 9 *Atlantis* steamed away from pier 18, and with a pilot guiding her through the early-morning traffic she moved into the Pacific Ocean. It was a gray, heavily overcast day, and the cranes and warehouses that outlined the port quickly faded from view in the haze. Captain Bray moved restlessly from his cabin to the wheelhouse and back again. *Atlantis* was now "on the other side of the gate," as the seamen said, and to Bray, who had hardly been farther than Cuba on the vessel before, it seemed a long way from Cape Cod. Within the hour Bray got the feeling that he was going to have one of his attacks, and although he tried grimly to argue himself into a calmer frame of mind, a sense of panic rose within him and to the surprise of all he abruptly ordered the ship turned around. Back to Balboa she steamed. The captain was met at the dock by an ambulance, and as he disappeared along the bumpy streets on his way to a hospital, the twenty-nine-year-old first mate, Dick Colburn, assumed command. Over crackling lines a call was put through to Woods Hole, Colburn was given the go-ahead, and at last *Atlantis* steamed resolutely into the headwinds and rain squalls of the Pacific and did not turn back.

A week later, after crossing the equator at two o'clock one raw, dark morning, *Atlantis* put into the port of Talara, Peru. Here the chief scientist, Henry Stetson, a friend of his from Harvard, and two Peruvian geologists joined the ship, and here the work of the expedition began. The ship left Talara on November 15. While still within sight of the dry mountains that rise behind the coastal plain, she hove to. A Stetson corer was swung over the side. Down it went, barrel, weight, and guide vanes, through the long Pacific swells, and up it came thirty-five minutes later covered with a stinking mud of mottled reds and grays. Again at 7:45 P.M., 9:00 P.M., and 11:00 P.M. the corer was sent down to the sea floor. Between stations the ship steamed along to the west, making a section across the northern portion of the Peru–Chile Trench. For two more days cores were taken every three to fifteen miles.

On November 18 *Atlantis* changed course and steamed southeastward against the Humboldt Current and parallel to the trench. Then a second section was made across a deeper portion of the trench as the vessel turned eastward toward Callao, Lima's seaport. Fourteen cores were taken on this profile, and along its entire length the helical wire in the echo sounder's new Alden recorder burned what looked like brown iron stains in a sweeping pattern of hills and hummocks.

Two days before Thanksgiving the lookout on *Atlantis* raised Palaminos Light, and early the next morning before the sun climbed over the rough, scrub-covered mountains, a pilot boat came bouncing out of the harbor to meet the ship. On it was Captain Bray. Apologizing, he explained that all his tests had again been negative.

As the pilot conned the ship between the curved arms of the breakwater and into the inner harbor, Bray learned from Colburn that the coring and sounding had been going extremely well, but that Henry Stetson was confined to his bunk with a painful ear infection. The chief scientist had declined Colburn's offers of medical aid, preferring to remain alone in his cabin and treat the infection with pills he had received in Woods Hole.

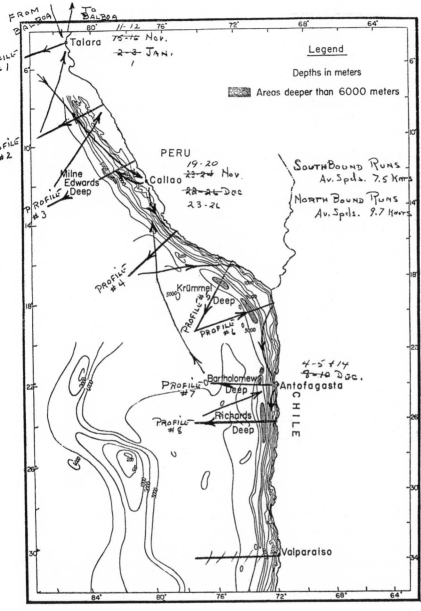

A rough sketch of the cruise track of *Atlantis* off Chile and Peru. Dozens of such charts were printed before an expedition, and as plans changed, the alterations were recorded. *(Courtesy Woods Hole Oceanographic Institution.)*

243

On November 25 *Atlantis* moved out of Callao through a low haze, and for a day steamed southeastward along the coast of Peru. On the twenty-sixth she turned abruptly west and the corer was swung over the side from the A-frame on the ship's starboard side. *Atlantis* was again moving across the Humbolt Current, and as she came on station the vessel became a part of a long, undulating line of shiny black cormorants. Thousands of birds sat in a great elongated patch on the gray water, and hundreds more flew above. The water was fairly boiling with anchovies, and all around the ship the cormorants were upending, waving their bright-red feet in the air as they dived for the fish. To right and left the birds bobbed back up to the surface, silver tails flapping from their beaks. The presence of *Atlantis* disturbed them not at all, and they rose and fell in noisy confusion as on the vessel the winch ground and grumbled, winding the corer in. When the 1,200-pound corer was raised to the peak of the A-frame, a line was passed around its muddy barrel, and when the barrel was drawn in and pulled aft, the corer lay horizontally just outboard of the ship's rail. Weight and barrel were then set carefully into brackets welded onto the ship's hull. The core barrel was unscrewed from the weight and lifted on deck. There, with one man pounding on one end of it and a second man pulling at the other end, the plastic liner with its three or four feet of chocolate-brown mud was extruded. The core was capped, and as the cormorants continued to dip and dive around the ketch, a new core liner was driven into the barrel and the barrel reattached to the weight.

John Zeigler, one of Stetson's friends and assistants, took the rough notes he had made on the latest core and went below to consult with Stetson. As usual, he found him lying in his bunk, his face gray and unshaven, his eyes watering uncomfortably. He didn't feel any better, Stetson answered in response to Zeigler's questions, and he was finding it increasingly difficult to sleep. Briefly Zeigler described the latest core and, ascertaining that there would be no change in plans, returned to the deck. Within an hour or two the ship hove to once more and the coring began again.

Atlantis continued to prescribe a series of giant zigzags across the trench, and at several points along the landward side her echo sounder revealed canyons and gullies reminiscent of those off Georges Bank that Stetson had explored in the 1930s. Whales and large squid were sighted frequently, sea birds flew constantly overhead, and from time to time silvery schools of fish surrounded the ship, reflecting the hazy sunlight that filtered through the clouds.

Coring on *Atlantis*. *(Courtesy Woods Hole Oceanographic Institution.)*

Every gray day the temperature rose into the low eighties, and every night it sank back into the upper sixties. It was a comfortable climate to work in, but not a cheerful one.

As the days wore on, one very much like another, Captain Bray became increasingly worried about Henry Stetson, and when the vessel turned offshore once more, he finally persuaded the chief scientist to accept several injections of penicillin.

Atlantis was crossing back and forth across a steep and fairly narrow portion of the Peru–Chile Trench now. The gash in the sea floor had already been followed for several hundred miles down the coast, and in this area off northern Chile it had a curiously flat bottom. Perhaps, thought Zeigler, turbidity currents were washing sediments down submarine canyons and slowly filling the trench.

On the last day of November, Captain Bray radioed for medical advice on Stetson and twice attempted to give him injections of penicillin with disposable syringes that annoyingly detached themselves from the needle as the injections were begun. Giving up on the syringes, Bray gave Stetson an antibiotic in capsule form. Although Stetson slept well and felt somewhat better, the captain ordered the ship to head straight for Antofagasta, Chile, two sailing days away. As he and Zeigler went off watch together at midnight and retired to the captain's cabin for a drink, as was their custom, Bray confided to the younger man that Stetson was a very sick man.

On the morning of December 3 Bray paced restlessly along the deck, watching the dark sky above the Andes turn to lighter shades of cloudy gray. He checked the night order book to be sure that Stetson had received his pills, and at breakfasttime hurried forward to ask the radio operator if he had received medical advice so urgently requested the evening before.

"No, sir," replied the operator, who was relatively new on the ship. He explained that to use direct voice communication to Antofagasta, he would have to wait till midmorning, when the station came on the air.

"Then use the key!" shouted Bray, his voice rising.

"But I might have to go through Chatham, Massachusetts, sir."

"I don't care if you have to go through *Moscow*, goddamn it!" Bray stamped out of the saloon.

The day wore on and *Atlantis* steamed steadily southward, making no stops for cores. In the afternoon Captain Bray retired to his cabin to give his aching legs a rest. In the galley Joe Lambert, the steward, fixed a cup of tea for Henry Stetson.

246

A few minutes before 4:00 P.M., Lambert made his way along the gently rocking companionway, passed through the lower lab, and stopped to knock on Stetson's door. Receiving no reply, he turned the brass knob and pushed the door open with his elbow. Face down on the floor before him lay Henry Stetson.

Lambert shouted for the captain, and as he ran from the cabin still clutching his tray, Zeigler rushed in. Quickly picking up Stetson's hand, he felt for the inert man's pulse. Finding none, he moved his fingers to the neck. Behind him, Bray's tall figure hurried through the narrow door.

"I turned him over for a quick examination," the captain wrote later. "His face was beginning to take on a bluish color suggesting asphyxia. We began artificial respiration immediately."

While the first mate, who had come in behind Bray, rhythmically leaned on Stetson's back and pulled forward on his limp shoulders, the captain ordered all speed for Antofagasta.

"Up jib, jumbo, and mizzen," he ordered. "Signals up."

Seamen from all three watches came running on deck. Sails were unfurled, hoisted, and set close-hauled to catch a gentle southerly breeze. The blue, white, and red W or "whiskey" signal flag was run up to the spreader.

"Require medical assistance," it stated mutely.

Below, oblivious of the sounds of running feet and rattling tackle, Dick Colburn worked over Stetson while Zeigler watched from the narrow passageway that led into the cabin.

At eight knots *Atlantis* rounded Point Tetas and crawled toward the small manmade harbor at Antofagasta. On her left, dry mountains rose from the sea, and far ahead, the breakwater framing the harbor's entrance lay like a gray line across the water. Light-colored buildings with red-tiled roofs rose hazily above the harbor on narrow terraces, and dozens of tall church spires attested to the relative wealth of this large mining town.

As the ship moved toward the harbor, a flashing signal blinked from her wheelhouse through the gray afternoon, and from her wireless set and voice radio came urgent requests for a doctor and a pilot. No one answered.

By 5:30 P.M., an hour and a half after Stetson had been found, the ketch hove to outside the breakwater. Her sails came rattling down, and with her signal light still flashing, her flag flying, and now her fine, mellow horn sounding a desperate SOS, she circled around and around outside the harbor entrance.

Henry Stetson showing what the cutting edge of a coring tube looks like before and after it hits rock. *(Jan Hahn photo.)*

248

Ten minutes went by, then twenty. In the chief scientist's cabin the first mate continued to work over Stetson. An hour went by. The lights in the city came on across the mountainside, and still there was no response.

Finally at 6:40 the Grace liner *Santa Barbara*, moored in the harbor, acknowledged the messages and relayed them to the Antofagasta Yacht Club. Within minutes a small skiff motored out to *Atlantis*. Even as it nosed alongside, Captain Bray was leaning over the rail. Waving aside the "Welcome to Antofagasta," he urged the man to find a doctor and a pilot. The skiff promptly put about, and some fifteen minutes later returned through the gathering darkness with a message from the port captain. No pilot was available, but *Atlantis* could proceed without one. A doctor would be waiting on the pier.

"1905. Under way. Entering Antofagasta harbor. . . . Captain has the con.

"1935. Tied up portside to dock."

Hardly had the docking lines been dropped over the bollards when two seamen on *Atlantis* set the ship's portable gangplank between the rail and the pier, where a doctor waited, bag in hand. Led by the captain, he hurried below. He ordered the artificial respiration stopped, made a brief examination, and pronounced Henry Stetson dead of a heart attack. Death, he said, had been immediate.

Late that night Stetson's body was carried up on deck and taken off the quiet ship. For the rest of the night *Atlantis* sat at the dock while around her moved the somnolent night life of a provincial port. A watchman passed with a lantern, an occasional sailor wandered back to his ship from a cantina, and the ubiquitous fog formed halos around the lights along the pier.

Henry Stetson had a relatively short career by many people's standards, but over the course of some twenty years he described in detail the sediments that cover the east and gulf coasts of the United States. Beginning his studies as a paleontologist at Harvard University's Museum of Comparative Zoology, he gradually shifted his concern from fossils embedded in marine sediments to the sediments themselves. Although he remained with the museum all his life, eventually becoming Alexander Agassiz Fellow in Oceanography, he accepted a second position as submarine geologist at the Oceanographic in the early 1930s. Soon he was working on submarine canyons, valleys, and other configurations all along the coast. Just before he died, at the age of fifty-five, he published a reevaluation and consolidation of his work: "Patterns of Deposition at the Continental

Margin." He was a close friend of Henry Bigelow's and Columbus Iselin's. In the best sense of the expression, he was one of the old guard.

Five days after Stetson's death, *Atlantis* steamed out of Antofagasta
to make two profiles across the deepest and most precipitous portion
of the Peru–Chile Trench. Parker Trask, who had taken over as chief
scientist, began the coring immediately, and when the vessel was
forty miles off the coast, a sounding almost five miles deep, a new
record for the eastern South Pacific, was made. As the ship continued
westward, the wall of the trench rose 1,500 feet in a quarter of a
mile. *Atlantis* had traced this narrow trench for nearly five hundred
miles, and from its deepest portions came essentially the same kinds
of sticky, dark-brown clays and olive-gray muds that had been taken
from the shallower regions.

As evening fell on the second or third day out, Bray and several
geologists gathered for drinks in the captain's cabin, and in the galley
the steward uncorked a second bottle of Chilean brandy to steady his
hand as he trimmed the meat. Work on the ship herself had stopped
for the day, and several of the seamen not on watch lounged on the
chartroom roof and watched the birds fly landward in the last light of
day. The sadness over Stetson's death was passing, and only a few
men continued to constrain their cursing or lower their voices, and
then largely out of respect for the feelings of the late chief scientist's
personal friends.

Later that night, after the first mate had had one of his rare suc-
cesses in obtaining a star sight in this region of perpetual overcast, a
light in a waterproof casing was let over the stern. Captain Bray was
especially interested in this form of night fishing. As the first curious
fish darted hesitantly in and out of the light, he helped one of the
crew bait a large steel hook and throw it over the fantail. Within ten
or fifteen minutes a small crowd of men had gathered in the narrow
space behind the wheelhouse to watch the dozens of fish and squid.
As the animals milled around the light, a squid darted into the center
of the circle, embraced the bulb in its housing, and abruptly shot off
into the darkness. Shadows of much larger animals seemed to move
along the outer edge of the illumination.

Suddenly a shout went up from the sailor holding the line as his
hook and bait were taken. Nearly swept off his feet by a flood of
advice and encouragement, he reeled in his line as best he could, and
there in the light of the underwater bulb writhed the most enormous
squid that any of the men had ever seen. Arms fully ten feet long ran

twisting up the fishing line like a mass of intestines, and the squid's body with its huge flat eyes was as large as a man's torso.

"Gaff it with the boathook!" shouted Bray, falling over men and lines as he ran forward to get the hook.

"Bring a dip net!" someone yelled.

As the commotion grew, men appeared from belowdecks dressed only in skivies or a pair of trousers and rushed to the fantail to join the struggle. Elbowing his way through the crowd, Bray arrived with the boathook, and in the shadowy darkness began lunging at the squid. With each mighty stab a bit of rubbery flesh flew into the air, but the hook would not sink in.

"Squid stew!" shouted Joe Lambert, who came wavering along the periphery of the battle with fumes of Chilean brandy wreathing his bald head and a ten-inch frying pan waving loosely from one hand. "I am going to make the best damn squid-and-spagetti stew you ever ate."

The huge squid twisted and turned, thrashing the water into foam. Then, with a series of violent convulsions, the animal tore itself loose and was gone. Nothing remained but a few small fish swimming in and out of the light. Reluctantly, fishing line and boathook were put away for the night, but for the rest of the watch Lambert roamed the deck with his frying pan.

"Squid stew," he was heard to mutter from time to time as he peered this way and that through his thick glasses, still hoping to find something to cook.

After running a box around the deepest portion of the Peru–Chile Trench, *Atlantis* returned briefly to Antofagasta, then headed north for Callao. In addition to the corer — several cores were put in the ship's freezer so that the hydrocarbons in the sediment could be measured later — both the BT and the GEK were used on this leg of the voyage. It was hoped that the character and position of the Humboldt Current could be determined. On December 18 the hundredth core of the expedition was taken, and on the following day the ship steamed into the lee of Callao's oil tanks and grain elevators and tied up for the Christmas holidays.

Because cruise 221 had extended over Thanksgiving and would continue, as had rarely happened before, over Christmas and New Year's, the Institution gave the men a trip to a large resort in the Andes.

Upon their return they had "a bang-up Christmas dinner" on the ship, wrote Captain Bray, "complete with songs sung by the pistol-toting dock guard, who gave me a few anxious moments when he wanted to shoot through the overhead."

Captain Bray had just received a letter from John Pike, and al-

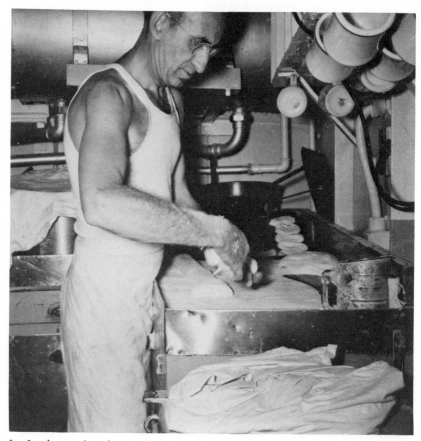

Joe Lambert making biscuits for thirty. When recovering from his "rum sickness" he would sometimes bake compulsively. Dozens of men still remember the smell and taste of his homemade bread, doughnuts, brownies, and sticky buns. He liked to feed people. *(Jan Hahn photo.)*

though it mostly described Woods Hole's sadness over the death of Henry Stetson, it also contained news of the Institution's attempts to obtain a 125-foot Coast Guard cutter. This was not the replacement for *Atlantis* that Brackett Hersey and Allyn Vine were after — the two were still maneuvering to acquire a large naval vessel — but the cutter would enhance the Institution's seagoing capabilities a great deal.

"We still do not know whether we own the *Crawford* or whether we will have to turn her back to the Coast Guard," Pike informed Bray. "We received the vessel due to our status as an educational institution . . . however, one of the higher-ups in Washinton has concluded that in his opinion we are not." (WHOI did not have a formal schedule of courses at that time.)

"We have attempted to prove conclusively to the man that we are. . . . Meanwhile, the *Crawford* just sits, and the scientists scheme. If for no other reason, the *Crawford* has been a good addition. . . . [She] functions as a diversion for our vast corps of scientific planners."

As *Atlantis* sat in Callao, five of the six geologists left the ship, and when the equipment that was to be used by the remaining man was lost by the airlines, the schedule of scientific work dwindled to nothing.

"Trip to Canal uneventful, with the exception of strong northerly winds," wrote Bray.

With the work of the expedition completed, *Atlantis* steamed through the Panama Canal and beat her way back into the familiar Atlantic, deadheading as she went. Of the ninety days of the expedition, forty-nine were spent in this fashion. Such inefficient use of ship time, which had so infuriated Maurice Ewing when he had worked in Woods Hole, was not the rule at the Institution, but periods of disuse were beginning to creep into the ketch's schedule. At any given moment there were usually several scientists with enough time and money to use the ship, but occasionally everyone seemed to fall in step, and then either everyone wanted to use *Atlantis* at the same time or no one wanted to use her at all.

"Got a gale off Hatteras," wrote Bray as the ketch sailed northward. "Dropped sails on a hunch at 1630 and at 1640 a front passed over with a change in wind direction that would probably have taken the sticks out."

The vessel sailed on up the coast, past the offing of the Chesapeake and past the low, gray outline of the Delmarva Peninsula. All feeling of spring was gone now, and by the time the ship steamed along the wintry hills of New Jersey, the days had taken on the dreary tone of a late-winter afternoon. Beams of deceptively warm-looking sunlight occasionally came in from the west through the wardroom portholes and rolled up and down the opposite bulkhead in time to the ship's motions. In the galley Joe Lambert's radio played on and on.

"Arrived in Woods Hole on January 26," wrote the captain, still suffering from his undiagnosed pains, "and so into the hospital for three weeks."

The hundred cores collected on the Atlantis Marine Geological Expedition were trucked from Woods Hole to Harvard University and from there samples were distributed to scientists whose major interests did not, unfortunately, lie in the Peru–Chile Trench. Five years elapsed before Parker Trask, who had taken over as the expedition's chief scientist, finally published a "preliminary report" describing the general conditions of sedimentation. On the basis of an examination of half the cores (the others were never even sliced open) he reported that in most respects sedimentation off Chile and Peru was proceeding as it did off more familiar coasts. Material was being washed from the land and deposited at sea — coarse gravels and sands near the continent and progressively finer muds and clays farther offshore. Planktonic debris was settling through the water column and mixing with this terrigenous material. Trask noted, however, that the trench was unusual in that it received very little material from the arid coast and also in that its sediments were extremely rich in organic matter owing to the presence of the overlying Humboldt Current with its rich populations of plants and animals. This second distinction, taken in conjunction with a brief study of the hydrocarbons present in the three frozen cores, led Trask to suggest that the trench might in fact be an oil field in the making. Although Trask, who was affiliated with the University of California at Berkeley, promised that "more detailed accounts will be published . . . in other journals," he died before writing further on the South American sediments. His partially completed analyses were packed in cardboard cartons and given to his wife.

At Harvard, where the cores were stored, another geologist wrote several papers on the physical and chemical relationships among the minerals sampled, but was not interested in the more basic questions that Henry Stetson had hoped to consider, such as how and when the trenches were formed.

The topography of the trench was described by John Zeigler and two cruisemates, but again, with Stetson gone, the desire to study the data in detail was lacking. The depth records were thrown away.

254

At the close of cruise 221, *Atlantis* received her annual overhaul before heading south with *Bear*. Hersey's group used the vessels for three months, and after the two ships had raced back from Charleston, South Carolina, still more acoustic work was pursued during the summer. The schedule seemed normal enough, but there were small yet significant changes. For one thing, *Crawford* had been acquired and, at a cost comparable to the building of *Atlantis*, converted into a research vessel. For another, Admiral Smith was retiring as director of the Institution, and for a third, plans were being made around the world for an extensive series of coordinated explorations to be made within an international geophysical year (IGY). Just as Smith stepped down, oceanographers were beginning to realize that this project could involve sixty to eighty research vessels from twenty nations and could be of great importance to marine science.

As plans for the IGY unfolded, it became clear to Henry Bigelow, who as chairman of the Institution's board of trustees still offered advice from time to time, that 1956 was no time to replace Smith with a new director unfamiliar with the funding and organization of oceanographic research. An experienced person was needed, and while Bigelow's appointee cursed him and publicly stated that the IGY was the only reason he "reluctantly agreed to resume, temporarily, the administrative responsibility for our research program," Columbus Iselin moved back into the director's office.

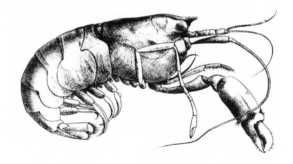

14 The International Geophysical Year

We lived in Indian summer and mistook it for spring.
— Bruce Catton

The job that confronted Columbus O'Donnell Iselin as he reorganized his books and files in the director's office after an absence of six years was the coordination of the Oceanographic Institution's research within a broad program set up by the International Geophysical Year. The IGY was to begin in July 1957 and end eighteen months later in December 1958. Unlike its predecessors, the First and Second Polar Years, it was to encompass investigations in all parts of the world.

The marine studies had three main objectives. The deep circulation of the sea was to be studied (with an eye to finding quiet abyssal pockets for the deposition of atomic wastes), sources of protein were to be evaluated, and the effects of the oceans on climate were to be measured. Iselin, fascinated since his college days by the Gulf Stream, was interested in the first.

The process by which scientists at WHOI selected their IGY projects was far more serendipitous than is generally imagined. There gradually evolved a plan to make hydrographic profiles at eight-degree intervals across both the North and South Atlantic, so that the circulation throughout an entire ocean basin could be studied for the first time. In addition, geologists and geophysicists at the Institution would take *Atlantis* into the Red Sea and Indian Ocean to run a cooperative seismic program with men from the Lamont Geological Observatory on *Vema*.

Several months before the IGY began, the converted Coast Guard cutter *Crawford* made the first three of some fourteen transatlantic profiles, and these early sections became the prototypes for IGY work. Echo soundings were made continuously and Nansen bottles with their reversing thermometers were let down as close to the sea floor as possible. The temperatures in the water column were recorded, as were salinities and concentrations of dissolved oxygen and phosphorus. As Iselin pointed out, until these sections were run, only a small number of casts had been made to depths greater than 10,000 feet, although most of the ocean is nearly twice as deep.

As the International Geophysical Year began, *Atlantis* was prepared for a spectacularly long cruise that would take her down the east coast of South America and across one of the broadest portions of the South Atlantic at 32° south. Although *Crawford* was already taking over cruises that previously would have been reserved for *Atlantis* (mainly because she was faster than the ketch and could keep moving in rough weather when *Atlantis* had to heave to), the cutter was too small to cross 4,000 miles of open ocean. Sail power still had at least one advantage. When welding slag trapped in an oil line forced the ketch to head for New York for repairs, however, and when a freakish series of storms slammed the ship around for eight days straight, smashing a wheelhouse window, damaging the whaleboats, and several times setting the ship on her beam ends, it became necessary to cancel two-thirds of the cruise. The portion that remained was the geophysical work in the Red Sea.

Lingering an extra day in Woods Hole to let a spring snowstorm blow inland, *Atlantis* left Great Harbor on April 3, 1958, and with as few stops as possible sailed for Alexandria, at the mouth of the river Nile. Scott Bray was still in command — still suffering from phlebitis and chest pains — and the ship was again deadheading with no scientific work to do. By early May *Atlantis* was steaming along the low, sandy coast of Tunisia, and on the evening of the eighth she slipped past the Ras el Tin Light and anchored for the night in the outer harbor of Alexandria. Near midnight the expedition's chief scientist, John Graham, motored out to the ketch with the harbor police.

When Captain Bray came on deck early the next morning, marveling at the flawless blue sky that arched overhead, a dozen skiffs and bumboats were already crowding around *Atlantis*, and the first Egyptian officials were chugging across the harbor in an open launch with their stamps, seals, stickers, lists, and forms. With customary hospitality Bray ushered them into his cabin, but as the day wore on and the customs and health inspectors continued to come panting in over the rail in their tight uniforms, and the knife sharpeners and the rug merchants in their baggy pajamas tried to follow, his patience wore thin.

"No boats are to be allowed to tie alongside or to hover alongside," he ordered, setting the crew to painting and red-leading all around the deck to repulse the infidel. "Any measure deemed necessary to keep them off outside shooting is permissible."

Despite his orders, confusion mounted. The robed and turbaned bumboatmen clustered around *Atlantis,* and with frequent invocations to Allah they tried to sell shoes, hats, camels, women, and pieces of the Cheops pyramid. Abruptly the noise around the ship subsided as one of the men on *Atlantis* took a picture of the bumboatmen with a Polaroid camera.

"After suitable mumbo-jumbo and magic passes," wrote one of the men, "the picture was produced. . . . [It was received] in a tizzy of sighs, moans, and exclamations, several of them kissing it with a most un-Moslem-like rapture."

With customs finally cleared, twelve tons of water pumped on board, and half the ship's company lost in the old quarter of Alexandria, Captain Bray propped his legs up on a chair and addressed his attention to a delicate matter. In a lengthy letter from Port Captain Pike he had just been warned that John Graham, the chief scientist, would try to persuade Bray to ignore the original plans limiting explosives work to waters fifteen or more miles off the Egyptian and Yemen coasts. Although only two years had elapsed since Egypt's nationalization of the Suez Canal had brought English, French, and finally United Nations forces into the Suez, and although sporadic fighting was still going on off Yemen, Graham wanted to work within five miles of the shore.

"Whereas you, yourself alone, will be solely responsible for irritations to non-Western sympathizers," wrote the ultraconservative Pike, "I rest assured that . . . [you will act] in a safe and circumspect manner."

In case his point was not perfectly clear, Pike went on to describe a predicament that *Bear* had just gotten into off Cuba. In mid-April *Bear, Crawford,* the Coast Guard vessel *Yamacraw,* and the U.S. submarine *Trumpetfish* had all been working with highly classified antisubmarine equipment off Cuba's south coast, not far from Fidel Castro's mountain stronghold. Late one afternoon several small planes came buzzing out from the island. Apparently, empty crates marked "TNT" and yellow plastic bags that were being used instead of balloons or cans to buoy up the charges had washed up on the beach and were beginning to suggest to Fulgencio Batista's army that the American vessels were not doing research but were trying to supply Castro with ammunition. The open-cockpit planes had been sent to investigate.

As the two-seaters swooped low over the ships that evening, the men on *Bear* and *Crawford* waved and shouted hello. Suddenly they saw that the man behind each pilot had a .50-caliber machine gun trained on the ships. In an instant the decks were cleared and all four vessels were steaming at full speed for Guantanamo Bay. The seas had been building all afternoon, however, and *Bear* was soon pitching and rolling over the black waves far behind the others.

On the high bridge, where the rolling was the worst, Shorty Mysona frowned and muttered as he watched a blip appear on his radar screen and move rapidly toward the ship. His eyes moved constantly from the radar to the dark night outside and back again, but whatever was closing in on him was running without lights.

"Dash-dot-dash-dot. Dot-dot-dot," flashed a light. "Who are you?"

Hanging onto the wallowing ship with one hand and working the blinker with the other, Mysona answered: "Research vessel *Bear*. Who are you?"

There was no response.

"Who are you?" Captain Mysona sent again, getting angrier by the minute. Still there was no response, and *Bear*, hove to now, yawed and heaved over the heavy seas.

Abruptly a carbon-arc searchlight burst upon *Bear*. At the edge of the blinding glare Mysona could discern the outlines of a large patrol boat and of a motor launch that was being put over her side.

Grabbing the radio, Mysona called for help. "Stand by me," he radioed the other research vessels. "I'm about to be boarded."

Crawford and *Yamacraw*, both loaded with explosives, received the message and kept right on going for Guantanamo Bay.

The launch from the patrol boat came climbing over the seas toward *Bear*, and soon Mysona could see that the navy men on board were armed with rifles and pistols. As they came alongside and prepared to board, Mysona stepped off the bridge and ordered them to stand off. Only one man would be allowed on his ship, he shouted angrily over the roar and splash of the water.

A young naval officer jumped agilely from the launch onto *Bear*'s deck, but before he could state his desires or intentions, Mysona, "as mad as a goddamned hornet," began hopping up and down and telling him that he had no manners, no common civility, no nothing!

At last the Cuban officer managed to convey, in his few words of English, a deep concern over the yellow plastic bags. Furthermore, *Bear* was less than ten miles from the coast, not the required twelve. He returned once more to the question of the yellow bags, and Mysona told him that his work was classified and would not be dis-

Bill Shields, left, and Hans Cook play cribbage in the shade of the whaleboats. *(Jan Hahn photo.)*

cussed. This response did not please the officer, and for a moment, as Mysona tried to translate what the Cuban was shouting to his comrades in the launch, he believed that *Bear* would actually be captured and forced into a Cuban port. Then from behind *Bear* came a bubbling and a splashing and a moment later the dark shape of *Trumpetfish*, her guns trained on the motor launch, came boiling through the black seas.

The naval officer leaped for his launch. Within minutes the patrol boat was again a mere blip on *Bear's* radar.

The moral of all this was to stay fifteen miles off Egypt and resist what Pike correctly assumed would be the chief scientist's tactless and single-minded attempts to do otherwise.

Dana Densmore, left, and Bob Munns enjoy Happy Hour on *Atlantis. (Jan Hahn photo.)*

As Bray got back on his feet and went to confront Graham on questions of national sovereignty at sea, the scientists and crew of *Atlantis* were already enjoying Alexandria, each according to his tastes. Bill Shields, who was now an engineer on *Atlantis* and had transferred his loud and unwavering allegiance from *Bear* to the ketch without a break in stride, impressed his new shipmates by selling Dana Densmore to an Arab for fifteen piasters. Densmore was one of the "water catchers" who was on board to make hydrographic casts. His hide was to be used as lampshade material, his beard as pillow stuffing. With his fifteen "disasters," as Shields insisted on calling the currency, the engineer swaggered into a city park and, pointing to what looked like two piles of white laundry, declared, "I'll take this one and that one." As he had frequently boasted, he blew his tubes in every port.

Meanwhile, the unflappable Densmore, who had lived most of his thirty-eight years along the shores of Massachusetts and had either worked on or repaired everything from sailing boats to a triple-screwed, three-engined World War I sub chaser, remained unaffected by the sale. With a relish he reserved for foreign ports, he and several shipmates took a horse-drawn buggy into the old part of the city and spent the entire evening in a café that he named the Valley of 10,000 Smokes.

"The café consisted of two walls of tattered carpeting and a tin roof," wrote Densmore in the log he had long kept at sea.

The whole quarter stumbled up to chat and smoke and cough. You can either unroll a cigarette and sprinkle in the hashish and reroll it (filter included!) or use, as most do, a hubble-bubble with a bowl, a beer can with the neck (for the water), and a two-foot reed stem. The stuff is cheap and plentiful and gives a good jag from the looks of the users. I didn't get any reaction to it, probably because of the beer. Good Moslems don't use alcohol, so hashish is the substitute. Evidently it affects the respiratory system somewhat, because after a pipe most of them cough and spit at length.

We left around midnight with great rounds of handshaking with Achmed, Abdul, and Mohammed, and clip-clopped back to Gate 6 at the Arsenal Basin. . . . Pop [who ran an Egyptian water taxi] was standing by with his fourteen-foot, lateen-rigged dinghy, and off we went. It's an hour's ride back to the ship at night with the light airs. Sometimes one of us rowed for a while, but mostly I sailed, everyone comfortable, the night warm and still.

On May 12 *Atlantis* left Alexandria and, skirting the edge of the Nile Delta, put in at Port Said. Two days later she fell in behind a convoy of twenty ships and began her transit of the Suez Canal. It was a clear, warm night when the line of lighted ships started south, and the four or five men who strolled along the ketch's deck could clearly see the canal's high, stone-clad banks passing silently on either side. At daybreak, as sunlight flowed over the eastern bank of the canal, the ships were still heading south, and although the canal widened as it passed through a lake or a by-pass, the convoy did not stop to let northbound traffic pass until it reached Great Bitter Lake, more than halfway to Suez. The temperature had already risen into the upper eighties and the sun felt uncomfortably hot to the men who stood on deck, taking photographs of the ships going north and of the small villages on the shore of the lake whose brown clay huts wavered in the heat. Densmore wrote:

Left Bitter Lake at noon and followed in solemn procession down the cut, past a revetted airfield with a dozen MIGs on it, past endless miles of kilo standards and big bollards for ships to moor to in emergency. . . . Occasionally there would be a dozen ragged tents and a work gang of men equipped with mattocks and basket, building the canal as they had done since the first one was cut between the Nile and Bitter Lake around 1900 B.C. . . . And just beyond one gang were two American drag lines doing the same work — monstrous things with 150-foot booms and enough diesel power to drive a battleship, biting out the hard red earth a truckload at a time.

Atlantis emerged from the 101-mile canal into the Gulf of Suez late in the afternoon, and after supper the third engineer broke a 6,000-mile string of bad luck and caught the first fish of cruise 242: a fifteen-pound bonito.

Although plankton tows and a few hydrographic stations were begun now as *Atlantis* steamed down the Red Sea, the real work of the cruise would not begin until late May, when the ketch would join *Vema* for the return trip to Alexandria. The major objective would be to make a dozen or more seismic refraction profiles in the Red Sea of the kind Ewing's group had been making since the late 1930s. Then as now, such profiles could provide a rough picture of the earth's crustal structure. Since very little oceanographic work had been done in the Red Sea, no one knew what sediments lay in its thousand-mile-long basin, how they were arranged, or why the narrow steep-sided sea had formed. The men from Lamont were more interested in answering these questions than were scientists from Woods Hole, for ever since Ewing had used *Atlantis* to explore the Mid-Atlantic Ridge, he and his colleagues had been fascinated by mid-ocean ridges and by the rifts that were often associated with them. In 1956, just a year before the IGY began, Ewing and Bruce Heezen had gone on record as saying that a worldwide system of interconnecting ridges existed. Now they hoped to examine this ridge and rift system further. Did the Mid-Atlantic Ridge indeed swing around Africa, join the Carlsberg Ridge in the Indian Ocean, then pass through the Gulf of Aden and into the Red Sea as a rift?

For the men from Woods Hole, who would turn over all seismic data to Lamont, the hydrographic casts made between profiles, the plankton tows, and the camera work were of more immediate interest.

With little work to do on the southern passage, then, *Atlantis* moved monotonously on through a shimmering furnace. Visions of passenger ships carrying pilgrims to Mecca wavered past during the day, and at night an endless succession of black, oily swells flowed quietly past the ship. Belowdecks the temperature was over 100; on deck, where BT slides turned white from the high humidity as soon as they were laquered, it was in the 90s; and in the stream of water that was constantly pumped across the deck and sprayed over the baking explosives, it was only 88.

"Slept on top of the lab last night," wrote Densmore, who, like many others, found it impossible to sleep below.

[I camped] under the awning over the main boom, mattress wedged between a collapsible rubber life raft and a huge spare anchor. With nothing over me it was fine sleeping, but still the sheet was soaked when I woke up for my watch. When I went below to get my toothbrush I stepped on Whitey Witzell, flaked out stark naked on a blanket on the cabin floor. There are bodies all over the decks, and Alfred John [second assistant engineer] has slung his hammock forward of the lab. Ought to buy one in Aden.

On May 21 *Atlantis* passed through the straits of Bab el Mandeb (the Gates of Affliction) and steamed into the Gulf of Aden. Turning east, she ran along the Arabian coast past mile upon mile of barren sand and chaotic eruptions of black volcanic rock. Not a plant or an animal could be seen in that blazing heat, and "I think," wrote Densmore, that even "the more insensitive of the crew were impressed by the bitterness of this land."

After a brief call at Aden, where, in addition to hammocks, the men bought turbans, fezes, and longhis, which gave the ketch a distinctly Eastern look, *Atlantis* proceeded into the Indian Ocean. On May 28 *Vema*'s three white masts were sighted on the horizon, and by afternoon the two ships were rolling and heaving together in a sloppy sea. After 1,600 pounds of TNT had been rafted to *Atlantis,* the first seismic profiles were begun. Advancing in turn, the two ships ran toward the island of Socotra, off Somaliland. Almost immediately a line of black squalls came sweeping across the gray seas from the west, interrupting the work and drenching the two ships with a cold, hard rain. For the next two days the two vessels headed into the weather, slamming into the seas and throwing spray all over the decks. No work could be done.

On the last day of May, after two months of sailing and two days of heaving to, the major work of cruise 242 at last got under way. Again setting off toward Socotra, *Atlantis* and *Vema* alternated between listening and making firing runs. On one of the first runs *Atlantis* started firing some thirty miles from the schooner. It was a fine tropical day, clear and hot, with a brisk wind blowing, and as had rarely happened on this cruise (and as never happened anymore on *Vema*), *Atlantis* swept along under full sail. A three-hundred-pound depth charge was heaved off the stern into her foaming wake, and a moment later there came a great slamming concussion against the hull followed by a low, watery rumble. Smaller and smaller charges were detonated as the ketch flew on toward *Vema* — fifty-five-pound charges that rattled the ship, then nine- and three-pounders that sounded like blows from a sledgehammer. By the end of the run whitecaps were hissing and tumbling over every sea and, wrote Densmore, "when we came roaring down on *Vema* in sheets of spray with everything drawing, they had every camera aboard burning film. . . . What a day! Everyone is grinning and cool and wet."

But when *Atlantis* became the listening ship, the heat again became a problem. With her engines shut down and with battery power sufficient only for a few essential lights and instruments, the ship became a steam bath. All fans and blowers were turned off and the reefers could not be opened for ice or cold drinks. As the temperature soared, the ship's company came sweating up on deck, and those who didn't have to slack the hydrophones or film the wiggle traces tried to doze on the chartroom roof or squeeze themselves into the meager shade of the awnings. Others sat under the stream of the deck hose for fifteen or twenty minutes at a stretch. Chief Scientist Graham and one of his assistants tried to make the periods of "silent ship" more bearable by gathering a supply of ice cubes when the ship was still under power and storing them in their thermos bottles. Later, when their shipmates sat on the quietly drifting ship drinking 85-degree beer, Graham and his student rattled the cubes in their tall frosty glasses of rum smash or martinis — and were hated for it.

Silent ship on *Atlantis* was frequently begun with an additional insult, delivered this time from the captain of *Vema*. Starting his firing runs with a flourish, as he loved to do, he would come barreling past the ketch with no more than twenty feet to spare between the two vessels. More than once he carried away some piece of the ketch's gear.

Densmore described the routine:

The ship-to-ship radio in the lab shouts, "*Vema* to *Atlantis* — the next shot will be number such-and-such, a 96-pounder with a fuse so long, approximate burning time 60 seconds." Then comes the one-minute warning and shouts to "stand by the hydrophones!" The radio sings out, "Over the side, over the side, the charge is over the side," and someone howls, "Slack!" and begins counting seconds with a stopwatch. On deck the two slackers are furiously paying out thick rubber-covered cable so the buoyed hydrophones are sinking neutrally [without dragging through the water] and not picking up any extraneous water noise. Then the radio picks up the growling double report behind *Vema* and gives three hoarse triumphant beeps and the slackers start to haul back the phones. Later, great dripping wads of film come out of the lab covered with cabalistic squiggles of refracted sound waves and join others soaking in huge cans of sea water.

After a week of shooting and listening, sailing and sweating, the two ships put into Aden for water and explosives, *Atlantis* taking on more than thirteen tons of TNT and depth charges. For the passage north through the Red Sea both seismic and hydrographic work was scheduled, and even without the extra strain that the heat imposed, the anticipated routine would be exhausting. On June 10 the vessels left Aden's sheltered harbor, with its narrow dirt alleyways, black-robed women, and enormous population of goats, and headed for the Gates of Affliction. Twelve hydrographic stations were to be made in quick succession to study the tongue of warm salt water that flowed out of the Red Sea and into the Gulf of Aden. While *Vema* took cores and used her magnetometer, the ketch hove to every two hours for two days and nights while Densmore and Conrad Neumann sent Nansen bottles just as deep as they dared. Between trips to the echo sounder to check the depth, a whole string of instruments was bent onto the hydrographic wire and, like "WHOI Bob," the antihero of oceanography whom Densmore had invented as he crossed the Atlantic,

> He hung each thing on that straining string
> That science had invented;
> "Oh, man alive!" he cried. "Down five!"
> He was a man demented.
>
> He had a tool for measuring cool
> And another for telling hot,
> A 3-D gear with a sonic ear
> For pictures on the spot.

So rigged, the "charm bracelet" was sent down, and while the thermometers were coming into equilibrium with the surrounding waters — 88°F. on the surface, 70°F. at the bottom — Densmore and

"Then the powers above looked down with love/ On Bob, for they could see/ That here was a man who could think and plan/ And unscrew the inscrutable sea." (*Conrad Neumann.*)

Neumann rewarded themselves with a tall rum drink with plenty of ice. Both read the thermometers when they returned to the surface, and any difference between the two readings was attributed to "Barbados parallax." The temperatures and a half-dozen other data were then entered in the limp pages of the logs, and later still, when the firing runs had resumed with *Vema*, the two water catchers sat for hours in the saloon making temperature corrections with their tables and slide rules. On an earlier cruise on *Crawford*, the mechanical problems of packing and stowing hundreds of water samples, from which salinity measurements would eventually come, had been avoided by the use of a salinometer, a new instrument that took advantage of the fact that sea water's ability to conduct electricity is dependent on its temperature and salinity. But as long as *Crawford* was working in the Atlantic, a more important area for the IGY work, the salinometer was not transferred to *Atlantis*.

Within the first week out of Aden minor calamities began to proliferate. *Vema* swept close past *Atlantis* and carried away her best hydrophone, a new form of heat rash broke out which covered the already uncomfortable men with small, painful blisters, and a headwind blew roughly from the northwest, slowing *Atlantis* to three knots and making work extremely difficult. The seismic work suffered, and the schedule was set back still further when Bray announced that because of headwinds, *Atlantis* would have to put into Port Sudan, halfway up the Red Sea, for additional fuel.

Fueling was duly accomplished in that sunbaked port, where handsome black Sudanese dockworkers carried oil aboard the ship barrel by barrel. On June 21 *Atlantis* left the port to rejoin *Vema*, and the excitement for the day was unwittingly provided by Dr. Graham, who returned to the ship with one hand bandaged and a bottle of slimy gray antirabies serum clasped in the other. He had, he said, reached through a fence to pat a dog and been bitten on the finger. Since it could not be ascertained when (or if) the mutt had been vaccinated against rabies, Graham would have to have an injection of the serum every day for two weeks. The first had been given in the hospital at Port Sudan; the remainder would be given by the man he liked least on the ship, Captain Bray.

On the following day, after Bray had given Graham the second 5-cc injection in his abdomen and had already decided (as legend has it) to brand his victim by making the shot marks read "SB," *Atlantis* hove to over one of several deep holes to make a hydrographic cast. The Nansen bottles went down about dark with a pinger and a camera to guide the way, but as had happened before, the exact location of the bottom got lost in the side echoes that came bouncing off the rough terrain, and the bottom four bottles got dumped on the sea floor. Remarkably, nothing was lost, and as the ship rolled easily over glassy black swells that sparkled with phosphorescence, Densmore and Neumann brought the bottles back over the rail and recorded the temperatures in the data logs. They noticed that in the lowermost bottles a queer reversal of the usual temperature pattern had occurred, and months later, when the water samples were analyzed in Woods Hole, it was found that the salinities were peculiar too. Instead of decreasing all the way to the bottom, both temperature and salinity increased markedly in the deep bottles. Question marks were put in the margins to flag these unusual data, and, as had happened on a Swedish expedition when a similar reversal had been found in the same area, the data were eventually incorporated into a cruise report without comment or interpretation. The findings were too bizarre to be taken seriously. (In 1963, however, an oceanographer at WHOI, convinced of the accuracy of Densmore's and Neumann's observations, diverted his ship to sample the area in greater detail. He conclusively showed that deep pools of extremely hot brine exist in the Red Sea — a unique occurrence, as far as is known today.)

For the next four or five days *Atlantis* and *Vema* made firing runs together as they moved steadily northward toward the Suez Canal. On June 27 the day's work was completed late in the evening, and as the temperature dropped into the low eighties and a warm breeze blew fitfully off the desert, Captain Bray left *Atlantis* in a rubber life raft to visit the schooner. He had bent his last hypodermic needle trying to get a shot into Graham, and he needed replacements. Unfortunately for the chief scientist, *Vema* carried only small syringes, which meant that Graham would have to have two shots instead of one each evening. Bray stayed on the schooner for several hours drinking with her officers and, thinking out loud of the greater number of shot marks that he would have to make on Graham, he is supposed to have contemplated expanding the tattoo to read "SOB."

Near midnight Captain Bray came paddling back over the swells to *Atlantis*. Tired and full of liquor, he pulled himself gingerly over the ship's rail and started below, stepping as lightly as he could on his swollen feet.

Graham abruptly appeared from the lab, where he had been awaiting the captain's return. He told Bray bluntly that he was not satisfied with the way the seismic work had been going. The routine would have to change.

Graham was so annoyed by the delays that he began to lecture

With all sail set *Atlantis* could tear along at better than 12 knots. *(Jan Hahn photo.)*

Bray on the efficient operation of a research vessel. His timing was atrocious. Not only was Bray feeling rocky, but he was already extremely angry at Graham. The chief scientist had taken the ship's navigation charts of the Suez without permission and sent them back to Woods Hole. For a minute the dazed captain swatted Graham's arguments aside like gnats, but when the old question of working in coastal waters was raised again, his head cleared and he blew up. He didn't care if *Vema*'s crazy Nova Scotian captain took his ship within ten or twelve miles of the shore to follow a goddamned rift, as Graham challenged, Bray wasn't taking *Atlantis* any closer than twenty — "twen-ty!" — miles of the Egyptian shore. Still in a rage, Bray repeated the order to the officer on watch and stumbled below.

The twenty-mile limit put an end to almost all useful work aboard

Atlantis, and as the ship moved northward up the middle of the Red Sea, Densmore and Neumann dawdled over two hydrographic stations to give Bray a chance to sleep his anger off and change his mind before the ship sailed out of position.

At 10:00 A.M. the decision could be put off no longer, and the bosun went below to ask permission to proceed to the next scheduled station. A curt nod from the captain and *Atlantis* swung toward the distant shore.

Bray emerged from his cabin about noon, and as the ketch moved quietly along under a hot blue sky he let himself down on the shady side of the deck lab to chat with Densmore and a technician, Whitey Witzell. Both men were fond of Bray, in spite of the disruptions that his temper and medical problems caused. The captain was so much a part of the ship, and his problems, like his strengths, were so typical of a seafaring life, that to disapprove of him would be to question the value of *Atlantis* and even the legitimacy of going to sea.

"The weirdest sort of misfits can fit smoothly into this pelagic community," Densmore wrote later, thinking of men with more obvious eccentricities than Bray's. And considering the reasons this should be so, he listed "pride of ship," which he felt held the crew together, and "respect for shipmate," which paradoxically ensured an inviolate privacy and right to be as you are to members of a group that lived right on top of each other. The regularity of the life was another factor, for in essence each man was told what to do and when to do it, what to eat and when to eat it, and all the rest — twenty-four hours a day.

As Densmore was rolling these thoughts around in his mind, Witzell straightened abruptly.

"What do I smell?" he asked.

The others sniffed and smelled it too.

"A capaciter burning out," he declared, and, followed closely by Bray, he ducked into the upper lab to take a look at the depth recorder.

Suddenly Bray jumped toward the lower lab.

"Sound the alarm!" he shouted, scrambling down the ladder with uncommon speed.

"Sound the alarm!" echoed Witzell and Densmore and followed him below.

There in the lower lab, on a hot plate, blazed a can full of paraffin. The flames from its mouth rose four feet in the air. As the ship's alarm was sounded and a devilish clanging rang through *Atlantis,* the entire ship's company came swarming up on deck and began to unlash the crates and haversacks of explosives. With the sound of running feet pounding overhead, Bray slapped a notebook over the paraffin, grabbed the sizzling can with a pair of pliers, and make his way cautiously up the ladder. Slowly he passed through the upper lab and emerged on deck. He was just approaching the rail when the notebook seemed to explode. Flaming paraffin engulfed his hand and arm and ran down the sides of the can. Bray dropped the can onto the deck; tongues of fire flowed along the planking. But two steps behind him was First Mate Dick Colburn with a carbon dioxide fire extinguisher, and with a hiss he put the fire out.

Bray was in excrutiating pain. He ordered the ship to steam straight for Suez, then quickly disappeared below with Colburn. The two dressed the burn as best they could, and Bray took a large dose of painkillers, pills with which he was already familiar.

By dinnertime Bray was feeling somewhat better, but the drugs had made him lightheaded and he acted almost drunk. When Graham appeared for his antirabies serum, he sized up the situation and announced that he would not take his shots.

"In my opinion Capt. Bray is not capable of judging whether or not I should receive a further injection of antirabies vaccine," scribbled the chief scientist in the ship's medical log. "The captain this date received a severe burn on his right hand and by this time has taken sufficient medicine for the relief of pain that I consider his judgment . . . impaired." He wanted, he went on to state abrasively, "the opinion of a competent medical authority."

That night *Atlantis* steamed into Suez, and as the morning sun rose above the tank farms and apartment buildings that roasted together on the flat sandspit, a doctor came aboard. He gave Graham his shot (and urged him to continue the series) and treated Bray's arm. For half an hour he and the captain remained closeted in the empty saloon, and the ship resounded with Bray's roars of pain.

"Please inform my family account of burns can't write easily. All is well," he cabled.

The major part of the work scheduled for cruise 242 was finished. *Atlantis* traveled back through the Suez Canal, and in Port Said and Alexandria most of the scientific staff left the ship and was replaced by a second group, which would undertake sonar research and other forms of acoustic work in the Mediterranean in the company of the Coast Guard vessel *Yamacraw*. In August the ketch would return to Woods Hole.

"Now that it's over I realize that I'm tired," wrote Densmore shortly before he left the ship. "I've been revolving around this bucket since early January, and two months of fighting the Indian Ocean, the Red Sea, and Graham have put the polish to it. I want . . . to see my family."

The results of cruise 242 were not long in coming out. The announcement that the Red Sea was indeed a part of the world rift system with a deep axial trough running down its southern portion was made at the first International Oceanographic Congress in New York City in the fall of 1959. The fifteen refraction profiles, the echo sounding, and the magnetic work all combined to show that beneath a layer of corals and unconsolidated sediments lay a thicker layer of carbonate rocks, and beneath those lay the continental basement. But in the central rift or trough, volcanic rocks seemed to underlie the sediments. More recent data confirm this picture and suggest that the Red Sea is a developing ocean in its early stages. It is slowly splitting apart along its axial trough, much as the Atlantic Ocean has done and continues to do.

A tentative answer to another of Ewing's major questions was provided by *Vema,* whose soundings during her passage through the Indian Ocean suggested that a previously undiscovered mid-ocean ridge extends southward from the southern end of the Carlsberg Ridge opposite Madagascar. On a subsequent cruise the new ridge was followed down the entire length of the Indian Ocean and around Africa to its juncture with the Mid-Atlantic Ridge.

The hydrographic data from the cruise took longer to come out. The really exciting discovery of the hot brines was hidden away in an unpublished "blue-cover report," and was exhumed for historical interest only after a later expedition had been made to the brine pools in 1963. A paper was published, however, on the circulation of the Red Sea. Since the 1930s it had been thought that strong prevailing winds were the most important force in the movement of the sea's excessively salty waters, but the work done on cruise 242 showed that the tremendous loss of water by evaporation and the resultant changes in density were more important than the winds.

Although cruise 242 was an IGY expedition, its results formed no part of the Oceanographic Institution's major contribution to the International Geophysical Year, the *Atlantic Ocean Atlas of Temperature and Salinity Profiles.* In other words, the ketch *Atlantis* was working outside the Institution's area of major scientific interest. In 1959 she finally made a transect of the South Atlantic at 32° south, reoccupying the German research vessel *Meteor*'s profile IV made in 1927 and showing that the ocean's temperature and salinity remain remarkable constant over thirty years. But it was *Crawford* that made the largest number of profiles and the greatest contribution to the atlas. *Atlantis* wasn't even second in importance in the IGY project. The research vessel that Brackett Hersey and Allyn Vine had been trying to pry loose from the navy for nearly ten years had finally arrived. With the advent of the 213-foot deep-sea rescue and salvage vessel *Chain,* the Woods Hole oceanographers with the most pressing work and the most ample grants no longer relied on *Atlantis.*

15 A Gradual Retirement

I never at any time regarded steamers as ships, nor worth a sailor's attention. When steam came in, beauty and seamanship departed.
— James William Holmes, *Voyaging*

Several months before *Chain* came churning into Woods Hole in a torrential November rain, Columbus Iselin stepped down as the Institution's director. He was replaced by Paul Fye, and with this replacement the modern era that Admiral Smith and Brackett Hersey had tried to initiate began with no equivocation.

For many Fye's arrival in 1958 was a welcome change. Having spent numerous years running scientific laboratories, first as resident supervisor of the Underwater Explosives Research Laboratory at Woods Hole during World War II and then as associate director for research at the Naval Ordnance Laboratory near Washington, D.C., Fye was an experienced administrator. He was more decisive than Iselin and more approachable than Smith. He listened carefully, he sought advice, and when, after a year or so, he decided the way he thought the Institution ought to move, he proclaimed himself in favor of expansion and professionalism. In his very first director's report he announced the establishment of committees for a building program, land acquisition, an educational program, and the design of new research vessels. Noting that *Atlantis* with her engine problems had been the weak link in the IGY projects and that the ship would need extensive repairs if her insurance rating were to be maintained, he planned to keep her in operation only until June of 1960. If logic prevailed, there would be nothing gradual about the ketch's retirement. Come June she would be put up for sale.

The cruises undertaken on *Atlantis* during what was supposed to be her last year of operation were largely unremarkable. She made two trips off the east coast to study currents and sediments — the final ones for Captain Bray — and then under her last American master,

Dick Colburn, she completed a series of expeditions of the Gulf of Venezuela. In 1953 the Creole Petroleum Company had invited scientists from the Oceanographic Institution to visit the oil fields of Lake Maracaibo and later to work in the lake and coastal waters. Four or five cruises had been made on *Atlantis* and data of particular interest had been brought back from the deep Cariaco Trench, just off the Venezuelan coast.

"The water in this great hole was found to be devoid of oxygen and

Paul Fye, the Institution's fourth director, retired in 1977 after 19 years. *(Courtesy Woods Hole Oceanographic Institution.)*

loaded with hydrogen sulfide," Admiral Smith had written after the first cruise. "The conditions are similar to those found in the Black Sea . . . [but] it is the first case where such conditions have been found to exist in a depression in the bottom of the open sea."

When *Atlantis* came north from Venezuela in March of 1960, plans were already under way for her participation in Gulf Stream '60, a four-ship survey of the Gulf Stream. In many respects the project was a more sophisticated version of Operation Cabot, launched ten

276

years before. The major piece of new equipment on Gulf Stream '60 was the Swallow float (named after its designer), a neutrally buoyant float that had been used since 1956 to follow deep currents. Each float was equipped with a sonic pinger and was weighted to sink to a certain depth. As it drifted with the deep water, its electronic pings could be followed by a ship on the surface. In addition to the Swallow floats, a plane would be used during Gulf Stream '60 both to track large surface buoys and to map warm streaks in the stream by means of a radiation thermometer. The use of Swallow floats introduced a new routine aboard *Atlantis, Crawford, Chain,* and the cooperating Coast Guard cutter *Evergreen.*

After the four vessels had each made nine deep profiles of the Gulf Stream, they set out from Bermuda to deploy and follow the submerged floats. For the men on *Atlantis* it seemed a queer operation compared to the usual method of occupying a fixed number of carefully spaced hydrographic stations. In early May, for example, a Swallow float, weighted to drift more than two and a half miles beneath the surface, was launched from the ship. As the slender aluminum tube was tossed over the side and began sinking through the water, hydrophones were lowered from the bow and stern of the ketch to pick up the sound from the float's pinger. Once the general drift of the buoy was ascertained and the ship's position precisely determined by Loran, the hydrophones were hauled in and the ship was driven ahead for fifteen or twenty minutes. Down went the hydrophones again, and the signal from the buoy was picked up. As the tracking proceeded, however, the signal was frequently lost, and *Atlantis* had to jog back and forth across a stretch of rough, gray Gulf Stream to relocate the faint underwater pings. At the same time strings of Nansen bottles with their reversing thermometers were being lowered to great depths in an attempt to bracket the track of the float with hydrographic stations.

With all this movement back and forth, the ship's engineers were commonly logging fifty or more changes in speed and direction in a single four-hour watch, and for all the men, asleep or awake, such frequent changes in the ship's motion were exhausting.

By May 13 *Atlantis* had followed the first Swallow float for more than two hundred miles in four and a half days. When its signal was permanently lost, a second and then a third float were dropped over the side. Even when the wind roared down from the north, opposing the flow of the stream and piling its gray waters into steep, choppy seas, *Atlantis* stayed with the floats, yawing and rolling.

After two months of arduous work, Gulf Stream '60 came to an end and the four ships headed for Woods Hole. *Atlantis* fell far behind, all but stopped by headwinds that blew day after day from the north.

Only two things could have brought on such a jinx, wrote Dana Densmore, who had frequently shipped out on *Atlantis* since 1957: someone either "threw a pair of shoes overboard or didn't pay his whore bill in the last port."

Atlantis banged and crashed through the seas, "booming like an iron canoe," and on June 15 crawled into Woods Hole in a dense fog. Most of the men on board were unusually happy following a conscientious attempt to abide by the customs regulations concerning liquor. They were allowed to bring in one open bottle apiece duty free. They drank the rest.

Like others of the Institution's "informalities," as breaches of etiquette or the law were sympathetically called, the long tradition of importing duty-free liquor had come to a sudden halt. The practice of sending *Asterias* or some other vessel out to meet *Atlantis* to "render all possible assistance" — that is, to offload liquor — was no longer approved by the administration (probably since Admiral Smith's time), so the cases of rum and whiskey came right to the dock under the noses of the customs inspectors. This disturbed the captains of *Atlantis*, *Chain*, and *Crawford*, for they were personally responsible for any infringements of the regulations, and they, not the seamen, would lose their papers and be fined if the illegal trafficking were discovered. Nevertheless, each man returning on the ships routinely brought back a case of rum or whiskey and sometimes a great deal more. Captains Bray and Mysona had tried to limit each man to six or eight bottles, which, although still illegal, were easier to hide. But when Bray put up a notice to this effect on *Atlantis*, and foolishly signed it, it was stolen from the bulletin board. Fearing blackmail or trouble with the customs officers, Bray had no choice but to demand an almost dry and perfectly legal ship for that particular cruise.

In spite of a few exceptions such as this, liquor had continued to flow into Woods Hole until March of 1960. In that month *Crawford* came into port on a Saturday afternoon and was cleared by inspectors from New Bedford. The next morning, however, as the usual process of unloading the contraband liquor began (timed to coincide with local church services), ten customs agents suddenly descended upon the ship. Seventy-five cases of liquor and at least two cars with their trunks full of rum were impounded. *Crawford*'s captain was fined more than $5,000 — a sum largely raised by his shipmates — and within days the investigation spread to the other ships in the fleet. Captain Bray was persuaded to admit to having illegally imported six cases of liquor rather than deny all culpability and risk a full investigation. He was fined $411, and he paid it himself. In spite of his agreement, several of the crew from *Atlantis* were called to testify in court. The most applauded performance was Joe Lambert's. The thin, nearsighted cook arrived in the Boston courtroom extremely drunk and swore he had consumed the twelve bottles of liquor he was known to have purchased in Bermuda in the four days it had taken *Atlantis* to sail to Woods Hole. His testimony could not be shaken, and his case was dismissed. Since March, then, *Atlantis, Chain, Crawford,* and the other WHOI vessels had followed the law to the letter, and even though Gulf Stream '60 had involved several trips to Bermuda, the ships came in nearly dry.

The data brought back from the four-ship survey were fascinating. When Swallow floats had first been used in the Gulf Stream in 1957, a majority of the deep floats had moved southward at a rate of some eight miles a day. A drift in this direction seemed to confirm theoretical studies that postulated the existence of a countercurrent beneath the north- and northeastward-flowing Gulf Stream. On Gulf Stream '60, however, the six floats launched by *Atlantis* in a more northeasterly portion of the stream indicated that the deep flow was moving in the same direction as the surface waters.

"The evidence from Gulf Stream '60 indicates that the Gulf Stream reaches to the bottom of the ocean," wrote Fritz Fuglister. And if the stream extended to the bottom, then the volume of water it transported could be twice the already phenomenal amount that was generally imagined.

"The [Gulf Stream] system can be likened to a mountain range," continued Fuglister, breaking significantly with the tradition that for a hundred years had called the Gulf Stream a river in the ocean. "The location of the whole seems obvious on a map of sufficiently large scale, but the boundaries of the features become indefinite when viewed in more detail."

Not only were the sides of the stream elusive: even its depth was a matter of conjecture. (Today it is generally believed that the Gulf Stream extends all the way to the sea floor, and although deep water can be found flowing south near its western edge, few speak of this flow as a countercurrent.)

Atlantis returned from Gulf Stream '60 in June, the month originally set for her retirement. But she was not laid up. Although Fye's committee on research vessels had already received a promise of $3 million from the National Science Foundation, "a grant which enables us to replace *Atlantis* with an equally fine ship," the new vessel was only a set of drawings. Scientists at Woods Hole needed a vessel from which to core, dredge, sample, and test. They couldn't all use *Chain*, whose schedule was dominated by Brackett Hersey's group, and there were simply too many projects for *Crawford* to handle. *Atlantis* was indispensable, and to the immense pleasure of all aboard her she was ordered to report to Munroe's shipyard for a general overhaul. According to the next director's report, those repairs would keep her going for another year.

Given this short reprieve, *Atlantis* embarked upon a series of cruises for chemists, geologists, and others, and in addition picked up the leftovers — outings for guests and associates, training cruises, and one- and two-day trips for the calibration of instruments. In November she headed south for a six-week hydrographic cruise under the direction of Val Worthington, a physical oceanographer who had been using the A-boat once or twice a year since the Med trip in 1947. Although the work was routine, the cruise itself was not, for, as Worthington claims, the spirit of Arvid Karlson sought revenge for a slight.

When Karlson died in 1955, his shipmates bought a clock and hung it in the ketch's saloon. On it was inscribed Karlson's name and the latitude and longitude where his ashes had been scattered. Worthington had always felt unhappy about not attending Karlson's funeral, and when on this particular hydrographic cruise he noticed that the ship's track would cross over the spot where Karlson rested, he decided to pour a libation on the waters in his honor. He would use a bottle of Karlson's favorite sugar cane brandy. The idea gave him considerable pleasure.

As *Atlantis* came back up the coast that December, moving gradually into rainy winter weather, Worthington forgot the libation. Almost immediately after the ship passed over Karlson's resting place, it was overtaken by a line of squalls. By early afternoon on December 12, *Atlantis* was rolling deeply in the trough of foam-streaked seas, and at 3:00 P.M. the engineer shouted up through the voice tube that the tail shaft — that Achilles' heel — was vibrating heavily. He suspected the shaft had broken somewhere in the stern tube. Captain Colburn ordered the ship hove to, but hardly had she swung into the wind when the broken shaft moved aft a few inches and jammed the ship's propeller into the rudder post. The vessel was now almost impossible to steer. Colburn had no choice but to remain hove to until the gale that had blown in behind the squalls began to abate, two days later. With a cold wind now blowing out of the northwest and a confused sea jostling *Atlantis* with an unpleasant motion, Colburn came up on deck to review the situation on his powerless and unmaneuverable vessel. He decided that he would have to dive down to the propeller, wrap some half-inch wire around it, and hope that the propeller could then be pulled away from the rudder when the wire was drawn forward through the hawse holes on either side of the ship. The ship would still be without power, but at least she could be sailed.

Shortly after breakfast, as a light rain fell from the overcast sky, Colburn emerged on deck in his bathing trunks, carrying parts of a shallow-water breathing device. He fitted the mask over his face, made sure that the air hose connecting him to a pump on deck was free, and slid quickly over the rail into the gray waves that swept along the side of the ketch. In a flurry of bubbles he disappeared beneath the overhanging stern. Behind him wire was payed out from a spool on deck. After what seemed like a long, cold time to the men crowded along the rail, Colburn surfaced. The ends of the wire were passed forward, and after the captain — the image of old Captain McMurray's down-Mainer who could go to sea for three weeks without going into a mental decline — came dripping and shivering back on deck, the wire was drawn tight and the propeller secured. Now that *Atlantis* was free again to sail, the mizzen- and headsails were hoisted and she resumed her northerly course for Woods Hole.

But on December 15 the powerless ship was becalmed, and on the sixteenth she was hit by a gale that blew out the jumbo and wrenched the propeller back against the rudder post. Again Colburn dived over the side and wrapped the propeller in wire, and again the maneuver was followed by a brief spell of fine sailing weather, a dead calm, and a gale. The pattern was too diabolical to be natural, and Worthington, now pouring the sugar cane brandy into his coffee, surmised that Karlson was teaching him a lesson.

On December 23 Captain Colburn ordered the mainsail set in an attempt to get *Atlantis* into a Boston shipyard in time for Christmas. The wrinkled main, which the captain claimed was continually weakened by smoke and gasses that blew from the ship's stacks into the damp furls of the canvas and caused a kind of chemical rot, presented a rare sight to the freighters that passed along the wintry coast. That night the wind rose only slightly, but the main blew out. The men on *Atlantis* spent Christmas Eve hove to on a lumpy sea waiting for a tow. A tug arrived at noon, and for the next day and a half the two vessels crawled toward Boston. At 10:00 P.M. on Christmas night they tied up at the deserted shipyard.

Although the expense of repairing the tail shaft was another black mark against *Atlantis,* her cruise schedule for the first half of 1961 was full and her retirement could not be advanced. Consequently, she left Woods Hole in January and made a long cruise to the Mediterranean to study the formation of bottom water — water that moves along the very bottom of an ocean basin and can be studied most easily in a small "model ocean" such as the Mediterranean.

Four months and a modest 138 stations later *Atlantis* was back in Woods Hole. Again she was due to retire, and again she was kept on through the summer, this time largely through the efforts of the Institution's biologists, who were beginning a more intensive study of the animals in the deep sea.

Since 1872, when HMS *Challenger* made a three-and-a-half-year circumnavigation and returned with the first extensive data on deep-sea life, the picture of the abyss had been one of a cold, black wasteland under tremendous pressure, sparsely populated by a very few kinds of worms and shellfish. Attempts to reevaluate these unproductive and all but inaccessible regions were begun in the 1950s. By the use of grab samplers that bit out measured chunks of the bottom, it was estimated that every square meter of the deep sea supported a population of about thirty tiny animals. This was more than had been suspected earlier, but still a small population compared to the coastal zone.

Then in 1961 biologists at the Oceanographic began using *Atlantis* to sample sea-floor communities between Martha's Vineyard and Bermuda. Using an anchor dredge that dug about four inches into the bottom as it was dragged across the sea floor, they sampled and resampled some dozen stations that ranged in depth from 100 meters to 5,000 meters — more than three miles deep. Each dredge haul was strained through a fine-meshed screen, preserved, and painstakingly examined. Even after the first cruise it was clear that there were many more animals living in the depths than anyone had imagined. And there were not only more animals (33 to 264 per square meter at depths greater than 4,000 meters) but more kinds of animals too. The most numerous were polychaete worms, crustaceans (particularly the deep-sea relatives of clams and oysters), and sipunculid worms, many barely visible to the naked eye. Less frequently snails, sponges, sea anemones, and brittle stars come up in the sediments.

After completing the summer's work and making several hydro-

An anchor dredge comes in over the rail and is about to be emptied into a wooden box. On the student cruises of 1963 and '64 the scientific work as well as the running of the ship was shared among students, scientists, and crew. *(Courtesy Woods Hole Oceanographic Institution.)*

graphic cruises along the Montauk Point–Bermuda transect, *Atlantis* set off in January of 1962 on another trip to the Mediterranean. As on the previous one, there were long free runs at either end and only half the usual number of scientists. Investigators from France, Yugoslavia, and Greece were invited to join the ship along the way.

In many respects the summer and fall of 1962 were the last regular season for *Atlantis*. Among her dozen short voyages, several were made for biologists and geologists and a couple for oceanographers getting ready for the International Indian Ocean Expedition. She carried the Institution's trustees to Newport, Rhode Island, to see the America's Cup races, and she was used to demonstrate a continuous seismic profiler to a group from the Massachusetts Institute of Technology. In September she carried the ashes of Scott Bray into Vineyard Sound. Just as Captain Bray had buried Arvid Karlson seven years before, now Shorty Mysona buried Bray.

When Bray had retired as master of *Atlantis* he had continued to work for the Institution as a photographer. He had lived alone on an old wooden boat. The captain was so thoroughly a seaman that when he died, certain of his shipmates believed the cause to be a curious wasting disease associated with a sailor's excessive exposure to the sun, rather than the circulatory problems that had bothered him for years.

In November of 1962, while the second *Atlantis* was being built at a shipyard in Baltimore, the old *Atlantis* made the last of what had become her traditional autumn hydrographic cruises. On November 28 she steamed quietly out of Great Harbor and, turning southwest, glided along the leafless shores of Naushon Island. It was a mild, partly overcast day, and the rain that morning had turned the island's fields to a golden brown and its thickets of woodbine and wild rose to an almost purplish black. Having no stations to make for several days, Dana Densmore and the two or three other members of the scientific staff moved leisurely around the ship securing cases of sample bottles and casually checking three enormous doughnut buoys that, with their 3,000-pound moorings, were secured on deck beneath a web of wire and rope.

Two days out of Woods Hole the inevitable storm came booming out of the northeast. A forty-knot wind screeched across the water, rain swept horizontally across the deck, and with the Gulf Stream pushing from the opposite direction the seas rose and broke in a confusion of spray and foam. *Atlantis* struggled on at half speed for several hours, the seven and a half tons of gear on her deck giving the ketch a ponderous, lurching gait. The wind strengthened toward midnight, and when Captain Colburn went out on deck to check the doughnut buoys, he was caught by a wall of water that poured in over the low rail and ran swirling back to sea through the tangle of lines and cables. He was very nearly washed overboard. Once back in the wheelhouse, he ordered the engine shut down. The ketch ran before the storm under bare poles.

"It's a good thing we *want* to go south," wrote Dana Densmore as rain rattled like birdshot on the ventilators and streams of water ran belowdecks as well as above.

When Densmore came up on the still wildly gyrating deck before dawn the next morning, the tumultuous seas were flashing with the reflected light of a violent electric storm. Thunder boomed ahead of the ship as the sky was ripped apart by six- and eight-pronged forks of lightning.

"So I put a set of Bach suites on the tape deck [originally put aboard to record whale noises]," Densmore wrote, "and propped myself up in the lab with Mattingly's *The Armada*. The violence and confusion of Elizabethan Europe nicely matched the howling tumult outside.

This business of foul weather existence is a queer one. Science is locked up in a lurching, rolling steam bath, dripped on from above, wet underfoot, with absolutely nothing to do for twenty-four hours every day. One can read just so long before the eyes (and possibly the stomach) revolt. One can nap fitfully on the bunk just so long before it becomes more restful to get up. Of course, there is no place to sit on this vessel except at the saloon table, where a moment's carelessness can cause a broken bone from the ponderous gimbeled table. Most of one's existence is passed . . . exchanging one small discomfort for another until — five days, a week, ten days [later] — port is reached or fine weather returns. It teaches patience certainly.

This particular northeaster blew itself out in only three days, and by the night of December 2, hydrographic station 6339 could proceed without worry or complaint. Under the deck lights on the gently rolling deck Densmore attached the yellow Nansen bottles to the hydrowire, and one by one they disappeared below the surface. The procedure, like the instruments, had hardly changed since the 1930s.

On December 3, one of the doughnut buoys with its heavy mooring and string of six current meters was laboriously launched, and on the fourth the last of the outward-bound hydrographic stations was made. The ketch was already approaching Nassau.

Although Densmore's work was done, he got up well before dawn on the last day out to stand a private watch for sentimental reasons.

"It was warm and lovely on deck with 'focking Wenus' newly risen . . . ," he wrote later. "The old Enterprise [diesel] . . . snored softly. Jimmy Cavanaugh was busy about his piloting; Johnny Crocker had the wheel."

After strolling for some minutes along the dark deck, Densmore climbed quietly down through the lab and made his way forward to the saloon for coffee.

In the galley Brooksy was talking to himself and dropping flares into the range in a cloud of foul smoke. I retreated on deck again under the swarming stars.

With the sunrise and Nassau sneaking over the horizon, I started [working] on the lab brass for the last time, rubbing up a buttery glow and a deal of sweat. Then a bucket of sweet water to clean the crusted salt off the noble teak brightwork around the companion[way] as the pilot came over the side.

By 8:00 A.M. *Atlantis* had entered the harbor at Nassau. A tug came chugging out through the clear waters to meet her, and, taking a line, dragged her unceremoniously in, stern first. Densmore left *Atlantis* the following day.

After the 1962 season it was downhill all the way for old *Atlantis*. There was no money for a long cruise to the Mediterranean or Caribbean that winter, and the vessel was laid up at the Riverside Yacht Yard in Jacksonville, Florida. She emerged in June 1963, and had not cleared port before misjudgment on the part of a bridge tender knocked twenty feet off the top of her mainmast as she steamed beneath a partially raised railroad span. Limping back to Woods Hole, she docked in what had already become in her absence the customary berth of the enormous *Atlantis II*. The 210-foot *Atlantis II* had been delivered in January and by June had already made six cruises, including a search for the sunken nuclear submarine *Thresher*.

We've given you your second *Atlantis*, the managers of the National Science Foundation and the Office of Naval Research were saying in effect; now for heaven's sake get rid of the first one.

But a certain segment of the staff at the Oceanographic — men who, like Val Worthington, believed that it made no sense to use a hugh ship when a small one could do the job; who, like biologist Howard Sanders, felt that the willingness and spirit of the men on *Atlantis* often made up for her mechanical disadvantages; who, like even the unsentimental Brackett Hersey, recognized that *Atlantis* was still an effective research vessel for scientists whose techniques had stayed pretty much the same — prevailed upon NSF one last time to

Atlantis II, the ketch's official replacement, was designed by the Bethlehem Steel Corporation and built at a cost of nearly $4 million. She carries 25 scientists and is run by a crew of 31. *(Courtesy Woods Hole Oceanographic Institution.)*

provide minimal funding for student cruises. The funds were indeed so meager that *Atlantis* had to sail shorthanded and could be operated only by having the students work as part-time crew.

Atlantis was laid up again for the winter of 1964, this time in a Boston shipyard, and although Captain Colburn's last sailing letter praised the summer training program and mentioned similar plans for the summer of '64, money for *Atlantis* had finally run out. For almost a year the Institution's newsletter carried a notice in its "Ship News" column — "RV *Atlantis*: in storage at Munroe's shipyard, Chelsea" — and then even this recognition of her existence disappeared.

Meanwhile, *Atlantis* fell rapidly into disrepair. Her decking and fittings cracked as she sat unprotected in rain and hot sun. Streaks of rust ran down every inch of her hull, and to anyone driving north out of Boston over the Mystic River bridge, she presented a distressing sight. The once-beautiful ketch was following the precipitous declines of so many famous research vessels. HMS *Challenger* had become a coal barge, the American *Carnegie* had exploded and burned, the first *E. W. Scripps* had been cut down for a garbage scow and the second sold to a movie company to be destroyed in a film.

There are few honorable ways of retiring an old research vessel and most are impractical. There was talk of making *Atlantis* into a museum, but this course was too costly; she might be kept on as a training vessel, but this notion was premature, since the Oceanographic was not yet a degree-granting institution; she could even be sailed out to sea and sunk — a variation of a Viking funeral. The more practical ways of disposing of *Atlantis* were too unpopular to be openly discussed, but the administration quietly cast around for a buyer and in 1964 had the good luck to find one. The Consejo Nacional de Investigaciones Científicas y Técnicas of Argentina, a state-run scientific and technological council, agreed to purchase the ketch for the token sum of $3,000 with stipulations that her name be changed and that she be sold back to the Oceanographic Institution if she were taken out of service as a research vessel. The sale was not formally completed until July of 1966, and that fall Captain Colburn and the ship's radio operator and electronics repairman, Tom Lyon, began commuting to Chelsea to help the Argentinians refit *Atlantis*.

The Argentine navy, whose men would operate the ketch, sent a crew of about fifteen led by Lieutenant Commander Ricardo Rennella. Many of his men had been specially chosen for their experience on sailing vessels, and this proved a wonderfully diplomatic move. By October, when the truncated mainmast was taken out of *Atlantis* and replaced by a tapered 128-foot hollow steel spar, the Woods Holers were unanimous in their admiration of the thoroughness with which the Argentinians were refitting the ketch. Such care had not been lavished on the vessel since the end of World War II.

Once the new mainmast had been fitted with a Jacob's ladder and painted buff, Captain Colburn was heard to remark, with obvious satisfaction, that it was barely distinguishable from real wood. Nor was the sail plan changed on *Atlantis,* nor were her brightwork and decking. The Consejo Nacional did have to replace the ship's wheel and bells, since they had been taken off as official souvenirs. The ketch's name was changed to *El Austral,* the Southerner.

In November 1966 notices appeared on bulletin boards throughout the Institution: "The A-boat will arrive for the last time at WHOI early

Last trip to Woods Hole. Captain Dick Colburn, left, sailed with *El Austral* and her new master, Ricardo Rennella, from Boston to Woods Hole in November 1966. *(Robert Brigham photo.)*

Tuesday afternoon, November 8. Ceremony will take place Wednesday morning . . . at 11:00 A.M. All come!"

Newly painted a glistening white, her first suit of Dacron sails bent on and ready, *El Austral* steamed into Woods Hole. Her crew dressed ship, and on the day of the ceremony flags flew all over the ketch, her brass and brightwork sparkled, and her company stood at attention in full uniform along her mahogany rail. She didn't look exactly like the old *Atlantis*, but she did look fine.

Work gradually stopped in all the laboratories and offices along Water Street, and by threes and fours men and women gathered on the dock. Scientists who had used *Atlantis* stood talking with younger men who had never been on her. Officers and crew from every ship in port gathered with their own to swap stories. Wives, secretaries, the grocery man, the drugstore clerk, kids too old or too young to be in school, and even the Woods Hole black dogs, a mixed breed whose numerous members form a part of every village gathering, all assembled on the dock that morning. All were admiring *Atlantis*.

There was an hour of speeches under a gray and chilly sky — a trading of compliments between the Argentine navy and the Institution's administration — but the people on the dock were left to piece together their own explanations of what the departure of *Atlantis* would mean to them. For some it signified the end of an adventurous period in oceanography when almost every cruise returned with a new species of fish, a new piece to fit into the Gulf Stream puzzle, a new interpretation of the structure of the sea floor. New techniques and instruments had been tried aboard *Atlantis* by the dozen, and the scientific papers describing the data they collected had been written at a furious rate. As Columbus Iselin had said years before, "We were skimming the cream."

To others, the departure of *Atlantis* meant the end of a particularly intimate relationship between scientists and the sea. A man could not work aboard *Atlantis* and ignore the sea, and there had been a feeling among members of the scientific community that this was a good thing. A closeness to the sea was an important ingredient in a man's research, perhaps because the open ocean had a way of adding common sense to visionary schemes and vision to overly practical ones.

But for the hundreds of men who had sailed on *Atlantis* as crew, the selling of the ketch affected more than their jobs. It meant the end of a rare and precious camaraderie, which was going out of Institution life and could not be duplicated aboard the newer vessels. On *Atlantis*, the close quarters, the sense of adventure, and above all the intense pride in the beautiful ship herself had bound crew and officers together. The ship had been a tight little island.

At noon a benediction was read over the ship, and with a sudden burst of applause the ceremony was over. The crowd dispersed along Water Street. On the ship herself a problem was discoverd in the generator and her sailing postponed.

RV *El Austral* finally left Woods Hole for Buenos Aires at 9:00 A.M. on November 11. Before she left, Captain Colburn had taken off her a yellowed and somewhat battered document that he had found in a drawer in her chartroom. It was a large, oddly shaped document, bound in red and white twine and stamped with a red wax seal. It was the original Danish bill of sale.

To all to whom these presents shall come, greetings: . . . Know ye, that Burmeister & Wain, Ltd . . . for and in consideration of the sum of one dollar ($1.00) and other valuable considerations . . . , does bargain and sell unto the said Woods Hole Oceanographic Institution . . . the whole of the scientific research vessel *Atlantis* . . . to have and to hold . . . forever.

Epilogue

When *El Austral* arrived in Buenos Aires, at the beginning of what was in those latitudes the summer season of 1967, it was discovered that the old ketch was still in need of some very fundamental repairs. In fact, to the disappointment of the Consejo Nacional, it was estimated that the work required to make her seaworthy by Argentinian standards would cost an additional 100 million pesos, or twice the originally estimated funds. This money was not available, and the vessel again sat idle, this time for more than two and a half years. Late in 1969 repairs were finally begun, and *El Austral*, having been moved to a naval base near Bahia Blanca, some four hundred miles to the south, embarked upon her sea trials.

In June of 1970 *El Austral* made her first oceanographic campaign, as they are called in Argentina, and with a group of graduate students and young scientists she cruised back and forth across the continental shelf off Puerto Belgrano. Her work included the collection of sediments, plankton, and water samples and the taking of soundings — a basic mixture that did not change much over the next two or three years.

The Argentine navy, which is still operating the vessel as planned, is not entirely happy with *El Austral*. She needs more modern equipment, especially a satellite navigation system, and she again needs extensive repairs. In 1978 she was restricted to coastal operations. Funds are only slowly becoming available for improvement and repair, and *El Austral* continues to function primarily as a training vessel. She makes five or six cruises each year, most of them of only a few weeks' duration, and spends roughly half the number of days at

An honor guard greets *El Austral* in Argentina. *(Courtesy Woods Hole Oceanographic Institution.)*

sea that she did as *Atlantis*. Still, she sails, she works, she is beautiful, and, as Yankee sailors sang for their famous schooner *Echo o' the Morn*:

> May she live to sail away
> To the boom of Judgment Day,
> When we hope to see her sailin'
> To the toot of Gabr'el's horn.

Sources

The major sources of information used for each chapter are listed below, and unless otherwise noted, are the property of the Woods Hole Oceanographic Institution. The deck log kept on *Atlantis* and the Institution's annual reports were invaluable aids for all chapters.

Chapter 1   Journal kept by Columbus Iselin on the maiden voyage of *Atlantis*.

Chapter 2   Journal kept by First Officer Thomas Kelley (privately owned).
"The Backus Factor," an unpublished article by Columbus Iselin.

Chapter 3   Journal kept by First Officer Thomas Kelley (privately owned).
Albert Eide Parr, "Report on Experimental Use of a Triangular Trawl for Bathypelagic Collecting," *Bulletin of the Bingham Oceanographic Collection* 4, no. 6 (1934).
Albert Eide Parr, "On the Longitudinal Variations in the Dynamic Elevation of the Surface of the Caribbean Current," *Bulletin of the Bingham Oceanographic Collection* 6, no. 2 (1937).
Letters exchanged between Albert Eide Parr and Columbus Iselin.
Letters from Captain Frederick McMurray to Columbus Iselin.

Chapter 4   Journal kept by First Officer Thomas Kelley (privately owned).
Henry Stetson, "Geology and Paleontology of the Georges Bank Canyons," *Bulletin of the Geological Society of America* 47 (1936).
Henry Stetson, *Oceanography*, fiftieth anniversary volume of the Geological Society of America (1941).
Columbus Iselin, "The Development of Our Conception of the Gulf Stream System," *Transactions of the American Geophysical Union* (1933).
George Clarke, "Dynamics of Production in a Marine Area," *Ecological Monographs* 16, no. 4 (1946).

Chapter 5   Columbus Iselin's notes for an autobiography.
Letters exchanged between Maurice Ewing and Columbus Iselin.
Journal kept by Allyn Vine (privately owned).

Sir Edward Bullard, "William Maurice Ewing, 1906–1974, "*Biographical Memoirs of Fellows of the Royal Society* 21 (1975).

Chapter 6   Journal kept by B. King Couper.

Letters from Captain Frederick McMurray to Columbus Iselin. Columbus Iselin's notes for an autobiography.

Journal kept by Charlie Wheeler (privately owned).

Newspaper articles from the *Falmouth Enterprise*.

Chapter 7   Letters from Captain Gil Oakley to John Churchill.

Fritz Fuglister and Columbus Iselin, "Some Recent Developments in the Study of the Gulf Stream," *Journal of Marine Research* 7, no. 3 (1948).

Chapter 8   Letters exchanged between Captain Adrian Lane and Gil Oakley.

William Wertenbaker, *The Floor of the Sea: Maurice Ewing and the Search to Understand the Earth*. Boston: Little, Brown, 1974.

David Ericson, Maurice Ewing, and Bruce Heezen, "Turbidity Currents and Sediments in the North Atlantic," *Bulletin of the American Association of Petroleum Geologists* 36, no. 3 (1952).

Chapter 9   Letters from Captain Adrian Lane to Gil Oakley.

Chapter 10 Letters exchanged between Captain Adrian Lane and Gil Oakley.

Fritz Fuglister and Val Worthington, "Some Results of a Multiple-Ship Survey of the Gulf Stream," *Tellus* 3, no. 1 (1951).

David Ericson, Maurice Ewing, and Bruce Heezen, "Deep-Sea Sands and Submarine Canyons," *Bulletin of the Geological Society of America* 62, no. 8 (1951).

Chapter 11 Letters exchanged between Captain Adrian Lane and John Pike.

Newspaper articles from the *Falmouth Enterprise*.

Chapter 12 Letters from Captain Scott Bray to Mary Bray and to John Pike.

J. Brackett Hersey, "Acoustic Instrumentation as a Tool in Oceanography," in *Oceanographic Instrumentation*, National Research Council Publication no. 309 (1954).

J. Brackett Hersey and Dick Backus, "New Evidence That Migrating Gas Bubbles, Probably the Swim-Bladders of Fish, Are Largely Responsible for Scattering Layers on the Continental Rise South of New England," *Deep-Sea Research* 1, no. 3 (1954).

Chapter 13 Letters exchanged between Captain Scott Bray and John Pike.

Parker Trask, "The *Atlantis* Marine Geological Expedition to Peru and Chile," *Nature* 177 (1956).

Parker Trask, "Sedimentation in a Modern Geosyncline off the Arid Coast of Peru and Northern Chile," *International Association of Sedimentology*, pt. 23 (1961).

Chapter 14 Journal kept by Dana Densmore (privately owned).

Egon Degens and David Ross, *Hot Brines and Recent Heavy Metal Deposits in the Red Sea*. New York: Springer-Verlag, 1969.

Walter Sullivan, *Assault on the Unknown: The International Geophysical Year*. New York: McGraw Hill, 1961.

C. L. Drake and R. N. Girdler, "A Geophysical Study of the Red Sea," *Geophysical Journal of the Royal Astronomical Society* 8, no. 5 (1964).

Chapter 15   Journal kept by Dana Densmore (privately owned).

Fritz Fuglister, *Gulf Stream '60*, vol. 1 of *Progress in Oceanography*. Pergamon Press, 1963.

Howard Sanders, R. R. Hessler, and George Hampson, "An Introduction to the Study of Deep-Sea Benthic Faunal Assemblages along the Gay Head–Bermuda Transect," *Deep-Sea Research* 12, no. 4 (1965).

Index

A number printed in italics indicates a page where the subject is illustrated.

ON ALMOST ANY WIND

Designed by R. E. Rosenbaum.
Composed by Jessamy Graphics, Inc.,
in 10 point Palatino, 2 points leaded,
with display lines in Palatino bold.
Printed offset by Thomson/Shore, Inc. on
Warren's Olde Style, 60 pound basis.
Bound by John H. Dekker & Sons, Inc.
in Joanna book cloth
and stamped in All Purpose foil.
Jacket design by Oliver Kline.
Watercolor of *Atlantis* on the jacket by Norman Fortier.
Jacket printed by Simpson/Milligan, Inc.

Library of Congress Cataloging in Publication Data
(For library cataloging purposes only)

Schlee, Susan.
  On almost any wind.

  Bibliography: p.
  Includes index.
  1. Oceanographic research. 2. Atlantis (Research vessel) 3. Woods Hole, Mass.
Marine Biological Laboratory. I. Title.
GC57.S34   551.4′6′072   78-58038
ISBN 0-8014-1160-2